Coral Reef Curiosities

Intrigue, Deception and Wonder on the Reef and Beyond

Chuck Weikert

Dayton Publishing LLC

Dayton Publishing LLC
Solana Beach, CA 92075
publisher@daytonpublishing.com
www.daytonpublishing.com

ISBN: 978-1732526532

Printed in U. S. A.

Publisher's Cataloging-In-Publication Data
(Prepared by The Donohue Group, Inc.)

Names: Weikert, Chuck, author.
Title: *Coral reef curiosities : intrigue, deception and wonder on the reef and beyond* / Chuck Weikert.
Description: Solana Beach, CA : Dayton Publishing LLC, [2020] | Includes bibliographical references and index. | **Summary:** "Richly illustrated, the book chronicles the secret lives of coral reef animals, and also reveals the connections between human cultures and this most diverse ecosystem on the planet, as expressed in art, science and literature from ancient times to today. Appealing to divers, snorkelers and armchair travelers, the author builds a fascination with reef creatures, to serve as a foundation for protecting them."—Provided by publisher.
Identifiers: ISBN 9781732526549 (hardcover) | ISBN 9781732526532 (paperback)
Subjects: LCSH: Coral reefs and islands—Effect of human beings on—Popular works. | Natural history—Popular works. | Coral reef animals—Popular works. | Coral reef ecology—Popular works.
Classification: LCC QH45.5 .W45 2020 | DDC 591.77/89--dc23

ABOUT THE COVER

Swirls of Blue Tang *(Acanthurus coeruleus)* and Schoolmaster Snapper *(Lutjanus apodus)* around a stand of Elkhorn Coral *(Acropora palmata)* in Virgin Islands National Park, St. John, U. S. Virgin Islands

ORIGINAL PHOTO: CAROLINE S. ROGERS

To Rosemary,
my forever snorkeling buddy

Contents

The Blade That Became a Book

My first exposure to reef life came not in crystalline Caribbean or South Pacific waters, but in a murky snorkel off the coast of Florida. There a toothsome barracuda half my size appeared out of the miasma, keeping close company as I paddled about its territory. I was at once nervous and astonished at how easy and joyful it was to observe another creature at such close range on its terms, not mine. I might not be able to fly with the birds — a long held dream — but I could float with the fishes. And so it was that the barracuda and all its buddies on the coral reef set the hook, reeling me in without a fight.

For me, exploring the reef as a snorkeler or scuba diver was a grand adventure, never knowing what marvelous creature would next reveal itself. Matching peculiar, wonderful names with these animals was half the fun and continually piqued my curiosity: bizarre labels that brought a smile just by saying them — gobies, blennies, frogfishes and Christmas-tree worms; names that commanded respect — stingray, scorpionfish, lionfish, fire coral; names that defy explanation — sweetlips, grunts and sea hares.

* * *

After many snorkels and scuba dives, and far from the reef, it was a knife handle that riveted my attention one cold, snowy morning in New York City. I couldn't take my eyes off it, even before I read the description, before I learned of its footnote in history. The object was a Bowie knife, a legendary blade I had used in its plastic form to lead many an imaginary charge as a child. And this one had belonged to the abolitionist John Brown. On a wet and chilly October night in 1859, Brown headed an insurrection that took over the armory at Harpers Ferry, Virginia. But he was captured by then Colonel Robert E. Lee and subsequently put to death, achieving a martyrdom that would kindle flames that consumed the nation a year and a half later.

That day in New York City the knife was part of an exhibit — Grant and Lee in War and Peace — at the New York Historical Society. Artifacts, maps and other displays interpreted how, individually and at their intersection, the two men altered the tapestry of American history. Their

SOURCE: VIRGINIA HISTORICAL SOCIETY

The "Hunter's Companion" Bowie knife confiscated from John Brown by Lieutenant J. E. B. Stuart

uniforms, Lee's saddle, and old photographs were interesting. But the mottled gold and brown pattern on the knife handle had the look of the familiar.

Smooth and sensuous tortoiseshell, without need of further polish, had by then caught my attention many times under much different circumstances. Beneath tropic seas I had seen its brilliant flash reflecting the sun's rays. Tortoiseshell practically begs to be stroked; ask any scuba diver who has fought the temptation.

The hilt of John Brown's knife could have come from no other source than the shell of a sea turtle, in particular a Hawksbill Sea Turtle. I wondered if the smooth, lustrous grip had been the selling point for Brown when he acquired the knife. And what was its provenance? Family heirloom, gift or just a simple purchase Brown had made from a village merchant?

I did some spur-of-the-web detective work and came up empty. But the web being the web, this led to that and I found myself entangled in a thicket of natural and cultural history. A perfect example: For 3,500 years, dating to Egypt's Queen Hatshepsut, Hawksbill shell has been fashioned into various articles — ornaments, jewelry, combs, cigarette cases, model ships, tea caddies — and knife handles. Two thousand years ago shell was traded as deep into the Americas as present-day Indiana and Ohio. Artisans, unable to resist the pliant nature of the raw material, established it as one of the planet's oldest luxury items.

It was all fascinating detail, but was there more to the story? This was "research rapture" — the delicious high felt when chasing elusive, illuminating threads down a bottomless rabbit hole — and I was all in.

A fragment of a wonderful reef creature had found its way not into a corner of a jewelry box but into the catalog of American history. On what beach, or in what net had its life intersected with the country's narrative? Thoughts of Brown's knife dogged me for the rest of that day. Sea turtle, knife handle, John Brown, a legendary event in the nation's history. There was something here — what was it? One hundred and sixty years after John Brown held the knife for the last time, its historical connection just begun, the luster, elegance and sheer beauty of the item had seduced someone else — me.

And then it hit me: The knife had brought me to a different way of thinking about reef creatures. I had never considered the realm of a coral reef beyond the magical diversity of life it harbored. But there were untold reef stories I had never thought about.

As the days rolled by, an idea germinated. In just this one exhibit, I had discovered something that joined the natural world of a reef creature with the world of humankind. Inspired to look beyond the typical coral reef guidebooks, the deeper I mined, the more astounded I was at the amazing breadth of this new field of inquiry. I found unexpected stories of dedicated individuals linked to reef creatures in strange and wonderful ways. I found reef creatures inhabiting poetry, art and science, living on in myth, folk tales and music. I began to research in earnest, following the diverse threads that interested me, and to write down the stories of reef creatures.

As the words flowed onto the pages, I came to understand that I was writing not just to entertain, but also to inspire, to encourage others to cherish what lies beneath tropic seas, to love the creatures of the coral reef. Because in the words of director, producer and cinematographer Louie Schwartzberg, "You're going to protect what you love." And as ocean temperatures rise, coral reefs can use some protecting.

I invite you to join me on a tour of the reef. We'll explore this splendid world prepared for adventure and ready for surprise, with no preconceived notion of what's about to come into view. The chapters of this book meander as you might wander the reef decked out in mask and fins, starting on the bottom with the coral itself, the sand around it and creatures that make their home on the reef or near it. We can move up in the water column to find the "hoverers," and farther up to the swimmers, like the Hawksbill Sea Turtle, that spend time in the open ocean. And then we can go back down, finally returning to the bottom.

You don't have to read the chapters in order, however. Read them in any order you like, as you choose your next subject from the animals listed on pages iv and v. And if you want a "deeper dive," the "Selected References" section that starts on page 112 offers resources for finding out more.

From the hundreds of species that could have found a home in this book, I have selected those that caught my fancy, just as on a real visit to the reef one might lose all sense of time, captured by an unusual sight. (A busy "cleaning station," for instance, with fishes patiently waiting their turn to be groomed, will do this to just about anyone!) I invite you to select any species, represented here or not, and peel back as many layers of natural and cultural history as you can to reveal stories unknown, stories of wonder, stories that I hope will do for you what stories of John Brown's knife did for me.

LUCA VAIME / SHUTTERSTOCK

A school of fusiliers and other fishes hover above the corals and other sedentary inhabitants of a coral reef.

MATTERS OF STYLE

Knowing about the style conventions used in this book may add to your enjoyment, and having this page to refer to later might prove helpful.

Species Names

Conventions for the names of species of animals vary from publication to publication. In this book the first letter of each word of a **common name of a single species** is capitalized; for example, "Green Moray Eel." So when you see a capitalized common name, you will know it refers to one species. A **common name used to refer to more than one species** is not capitalized; for instance, "moray eels."

The **scientific (Latin) names of species generally do not appear in the main text** of the chapters. This is so these names don't interrupt your reading as the story moves along. In general, scientific names appear only when the text discusses something about the name itself. **However, in photo captions,** when you're looking right at a picture of the creature in question and the text is short, the scientific name is often provided along with the common name, in case you would like to know both.

Singular or Plural?

The **singular form** is used for **either one individual animal** (for instance, "the small fish darted away") or for **several individuals of a single species** (for example, "a school of Bump-head Parrotfish." You can tell which from the context of the sentence.

The **plural form** — for instance, "fishes" or "porcupinefishes" — refers to **more than one species.**

Measurements

The standard used in science worldwide and adopted for other aspects of life in most places on earth is the **metric system.** So in this book, weights and measures are given in metric units. If you are one of the many people in the world who tend to think in the **English system** instead, here for your reference are some quick, approximate conversion factors:

- approximately 2.5 centimeters (cm) = 1 inch
- 1 meter (m) = approximately 39 inches, just a bit more than 3 feet, or 1 yard
- 1 kilometer (km) = approximately 0.6 mile
- 1 kilogram (kg) = 2.2 pounds
- approximately 30 grams (g) = 1 ounce

1

Sixty Degrees of Separation

We feel surprise when travelers tell us of the vast dimensions of the Pyramids and other great ruins, but how utterly insignificant are the greatest of these, when compared to these mountains of stone accumulated by the agency of various minute and tender animals.

— Charles Darwin, April 12, 1836 entry, *Journal and Remarks,* published in 1839*

Imagine the great naturalist writing those words, tucked in his cabin on board the HMS *Beagle* anchored off an uninhabited island deep in the Pacific. The unfocused reptilian and avian eyes of jarred specimens, collected from the Galapagos Islands and other far-flung destinations, gaze upon the man whose prose will one day shake the world. All of that would come much later. For the moment, Darwin puzzled over how it was that coral atolls formed in the middle of tropic seas. In *The Structure and Distribution of Coral Reefs,* published in 1842, he would "describe from my own observation and the work of others, the principal kinds of coral-reefs... and to explain the origin of their peculiar forms." His theories, tested by time, clarify not just how coral atolls establish themselves but how the formation of a coral reef is the production of billions upon billions of those various small animals — coral polyps.

* Republished in 1905 as *Voyage of the Beagle*

Charles Darwin's ship, in an engraving from Voyage of the Beagle*; adapted from R. T. Prichett's* HMS *Beagle* in the Straits of Magellan *(oil on canvas, c. 1890).*

The story of coral reefs began millions of years ago when a single-celled marine alga came into contact with a coral polyp, basically a tube equipped with a crown of tentacles used to sting prey. Freighted with chloroplasts that harness the sun's energy by combining sunlight, water and carbon dioxide into energy-rich food through the process of photosynthesis, the alga was a tempting partner for the polyp. The polyp extended an invitation the alga couldn't refuse and the two have lived happily together ever since, the algae inside the polyps' tissue, in symbiotic bliss most of the time. The algae, tongue-twistingly known as *zooxanthellae* (zo-zan-THELL-ee) benefit from the carbon dioxide and other waste chemicals emitted by their landlord the polyp, using these compounds for photosynthesis. In exchange for these wastes and for holding the algae in optimal sunlight and protecting them from passing eaters, the polyp gained a source of energy — carbohydrates produced by the algae's photosynthesis.

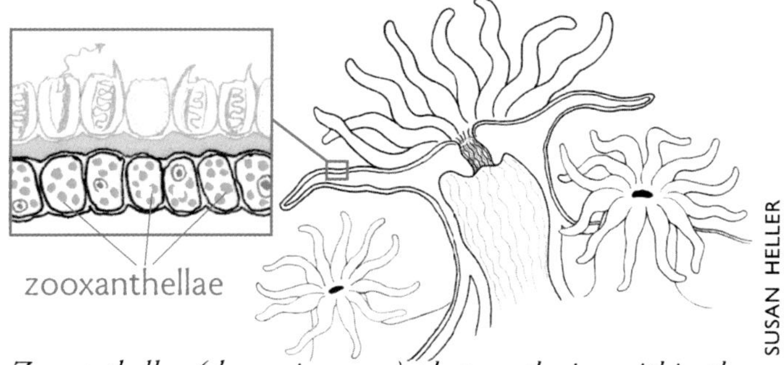

Zooxanthellae (shown in green) photosynthesize within the cells of coral polyps. An outer layer of cells (shown above the zooxanthellae) often includes stingers that protect the polyps.

With time, coral polyps build up a calcium carbonate outer skeleton, shaping massive coral reefs, some large enough to be seen from space. Located in a 60-degree band that extends roughly from 30 degrees south latitude to 30 degrees north, reefs add up to only about a tenth of 1 percent of planetary real estate, and yet coral reefs sustain millions of species of plants and animals, including a quarter of all marine fish species.

Three decades before Darwin set his quill to paper, English sea captain Matthew Flinders also had something to say about coral reefs for the folks back home. After spending weeks charting the Australian coastline, he must have found the Great Barrier Reef to be a pleasant, albeit hazardous diversion. Following a slog about the surface of the reef on foot, Flinders returned to his ship, the HMS *Investigator*, where his inner poet took the helm:

> *There were wheat sheaves, mushrooms, stags' horns, cabbage leaves, and a variety of other forms, equalling in beauty and excelling in grandeur the most favourite "parterre" of the curious florist.*

In his log Flinders made no attempt to account for the origins of coral. But ancient Greeks, never shy about such things, had their own ideas. For them, its origins were hitched to the hero Perseus. Perseus was selected by the gods to slay the nightmarish Medusa, dreaded

A school of grunts investigating staghorn corals

sorceress whose bad-hair day meant taming a scalp replete with hissing snakes. Glimpsing her was quite the lasting experience, as the unfortunate onlooker was immediately turned to stone!

Clever Perseus decapitated Medusa using his reflective shield to peer at her image so he could see where to strike without looking directly at her. Later, exhausted after dispatching some evil-doers with his new weapon of choice, he laid the magical head at the edge of the sea. When Medusa's blood changed seaweed into coral, it captured the attention of a group of sea nymphs, who had the brilliant idea to spirit the head down to the seabed. There, so one version of the story goes, Medusa continued to work her magic, positively this time, transforming rocks and plants into the world's first coral reefs.

No matter how captivating the legend, it pales in comparison to the experience a real live coral reef provides.

SOURCE: METROPOLITAN MUSEUM OF ART, OPEN ACCESS PROGRAM

Coast View with Perseus and the Origin of Coral *(Claude Lorrain, 1674). At the left nymphs gleefully dance around Medusa's severed head. At the right a helmeted Perseus cleanses his hands of her blood. Newly sprung from Medusa's headless body, the winged horse Pegasus seems poised for adventure. Beneath the water stretching to the horizon, a world of immeasurable beauty and complexity is about to be born.*

MATTHEW FLINDERS

Finding the name of the man who wrote so eloquently of coral reefs as he circumnavigated Australia is not difficult down under. Streets, stations, mountains, towns, an Australian river, a reef and a university are just a few of the ways the name of Flinders lives on. Up north, time and a 19th-century London railway station were long thought to have erased his final resting place. However, in early 2019 the captain's remains were identified at a construction site and were to be moved to a place more worthy of the famous navigator.

MARK RICHARDS

Sculptor Mark Richards's bronze statue commemorating Captain Matthew Flinders, with his sailing companion Trim the cat, resides at Euston Station in London.

When gliding down to a healthy coral reef, we are greeted by fins, tentacles and Picasso-esque distractions. Beyond those, we discover a diverse collection of what are essentially mouths angled toward sunlight, coral polyps that for most species emerge after sunset to feed on the sea's organic soup. Collectively, the polyps can form room-sized boulders of brain and star corals, pillars 3 meters tall, dense thickets of branching corals and a few species that rival the most stunning flower blossoms imaginable. The psychedelic colors of coral colonies — neon purples, luminescent greens, oranges and blues — shout for our attention.

Other attention grabbers are the colorful basslets, blennies, gobies and numerous other small fishes that dart among the corals. When threatened, anthias fold as if by magic into the reef or into the embrace of an anemone's tentacles. Snappers, sea turtles, parrotfishes and lobsters shelter beneath overhangs or in the reef's passageways. In the water column above the reef, thrilling transients — mantas, tunas and sharks — solo by, or visit in heart-stopping schools so dense as to swallow the sun's rays. This complex network of life eventually leads to the glorious and the grand — Goliath Grouper, Green Moray and Humphead Wrasse are but a few of the wondrous fishes that capture the imagination.

Below, within the shadowy passageways of a healthy reef, the coral remains its cryptic self, its curiosities and complexities mysterious and unknowable, a place where the writer Somerset Maugham described "fish swimming, like natives of the forest threading their familiar way through the jungle." The ancient Greeks tried to puzzle out coral reefs. So too various sea captains, naturalists and scientists who have gradually unlocked many of the reef's secrets only to find other mysteries to unravel.

PHOTOS: CAROLINE S. ROGERS

Some of the many forms taken by corals (clockwise from top left): Doughnut, Artichoke, or Disk Coral (Scolymia spp.); Boulder Brain Coral (Colpophyllia natans); Pillar Coral (Dendrogyra cylindrus); Smooth Flower Coral (Eusmilia fastigiata)

JANE GOULD / ALAMY

A "cloud" of Lyretail Anthias (Pseudanthias squamipinnis) hover above corals in the Red Sea, prepared to retreat to the safety of the reef if threatened.

2
Animal Blossoms

If the coral reef is a marine garden, then sea anemones are its elegant, blossoming flowers. Their tentacles, which range from ivory to pink, yellow and green, flutter and wave before the ebb and flow of water currents. Seemingly uncomplicated animals — solitary polyps attached to the sea bed — they are anything but simple once you get to know them.

CAROLINE S. ROGERS

An adult Queen Angelfish (Holacanthus ciliaris) "poses" next to a Giant Caribbean Sea Anemone (Condylactis gigantea).

There are some 1,000 species of anemones. They range from thumb-sized to double-arm-span in diameter and thrive in diverse marine habitats, ranging from tropical reefs to frigid polar waters. They live on coral reefs, in the sand around them, sometimes even in mud. They are collected by scientists in deep ocean canyons and in tidepools, objects of curiosity for children as well as naturalists. Underwater photographers count them as among the most patient and willing of models.

Sea anemones are members of the phylum Cnidaria (Nih-DARE-ee-ya; the "C" is silent). A group of organisms many thousands strong, it also includes the reef's foundational hard corals. Cnidarians have a single body cavity that serves as central headquarters for all of life's basic functions. If a morsel drifts into their maw, gulp it they will. In addition, they have devised a handy way of active food collection.

Welcome to the realm of one of the world's most voracious and efficient predators. Anemones are spirited diners that gather food in a fascinating if pernicious manner. Common to all cnidarians are specialized structures called *nematocysts* (ne-MAT-uh-sists). They give jellyfishes their sting, fire corals their burn, and the tentacles of some anemones their sticky feel. Each nematocyst contains a harpoonlike shaft complete with spring and poison sac attached. Like a twitchy medieval sentry waiting to launch its arrow, each nematocyst is primed to fire at the slightest touch.

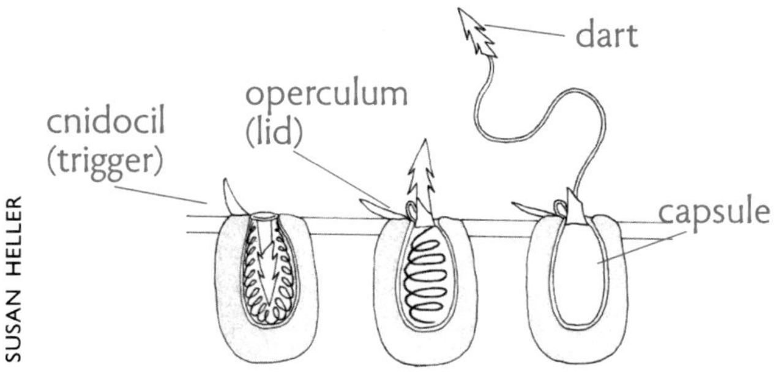

When a small fish or other prey brushes against a nematocyst's "trigger," a "lid" pops open, water enters the "capsule," and a coiled thread expands instantly, propelling a "dart" (in many species the dart is barbed).

Anemones dart their prey, small fishes as well as tiny *plankton* (organisms drifting in the sea), and shuffle these paralyzed creatures to their mouth. This feeding strategy earned them Scottish naturalist J. G. Dalyell's label as "devourers of whatever they can overpower." The stinging cells double as a perfect defense mechanism, and most predators know better than to chance the encounter.

Many sea anemones participate in *symbiosis,* or mutually beneficial relationships, on coral reefs. They put out the welcome mat for a diversity of crabs, shrimps and fishes that are unaffected by their nematocysts. Thanks to Nemo, beloved star of animated films, the best known of these symbionts are certain species of clownfishes. Nestled within the protective tentacles of their landlords, clownfishes hole up in these permanent, sheltered quarters. From here they forage. They also lay eggs among the tentacles, waging battle with anything that would threaten their offspring or their host.

A Giant Caribbean Sea Anemone (Condylactis gigantea) with a Spotted Cleaner Shrimp (Periclimenes yucatanicus)

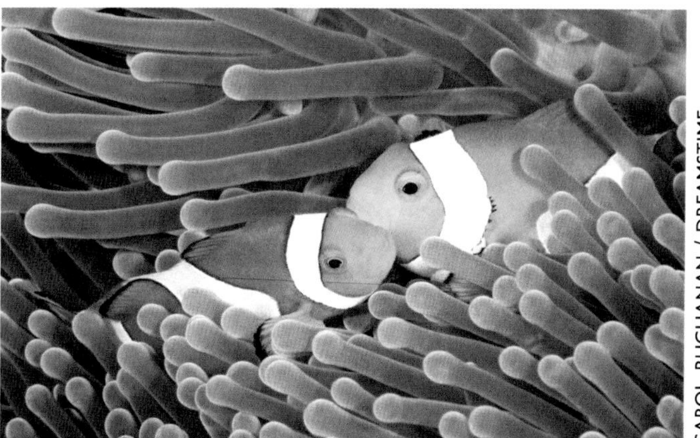

Possibly unnerved by the photographer's lens, a pair of Clown Anemonefish (Amphiprion ocellaris) tuck themselves in among an anemone's tentacles.

EVIDENCE OF ANEMONE EATERS

"... and finally, allow me to offer you some marmalade of sea anemone, equal to that from the tastiest fruits."

— Captain Nemo to Professor Arronax, in *Twenty Thousand Leagues Under the Sea* by Jules Verne, 1870

Anemones are capable of moving short distances but in general are true couch potatoes, content to hang out in one place for extended periods. Even so, a marine version of the "sharing economy" seems to exist, and alternative modes of transportation are available, in some cases whether the anemone likes it or not. For example, meet the diminutive, but somewhat feisty Hawaiian Boxer or Pom-pom Crab. Not content with the hard carapace nature provides for its protection, the crab totes about an insurance policy in the form of anemone "boxing gloves," brandishing them, armed with nematocysts, against approaching predators. In return, the anemone foot soldiers are rewarded with bits of food stirred up as the crab muddles about on its peregrinations.

If you're a Tricolor Anemone, then the Stareye Hermit Crab might be your jitney of choice. The crab taps the anemone as if to say, "Look, I've got room aboard, so if you'd like, climb on up." Crabs even joust with each other for the privilege of providing this service. In return for protecting the crab from foraging octopuses and the like, the anemone gets to dine on tidbits left over from Crabby's meals.

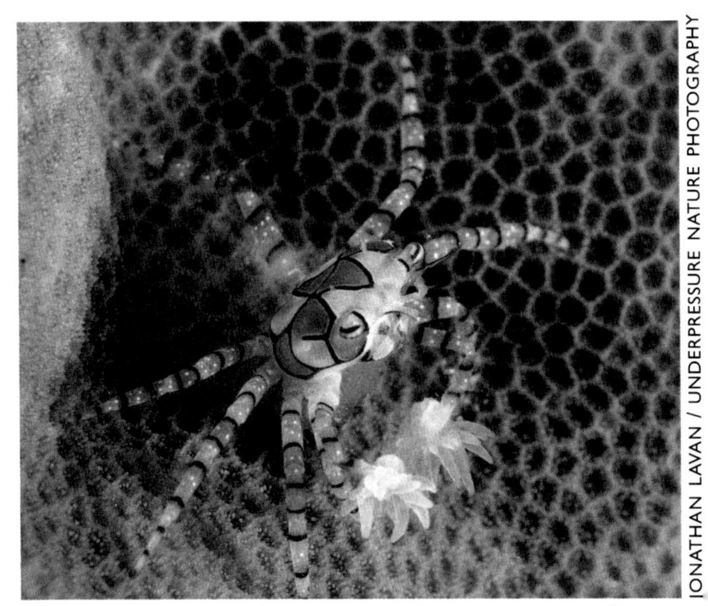

JONATHAN LAVAN / UNDERPRESSURE NATURE PHOTOGRAPHY

Laden with a mass of orange eggs, this Pom-pom Crab (Lybia edmondsoni), also known as a Hawaiian Boxer Crab, brandishes two Triactis producta anemones as a threat.

JOHN ANDERSON / DREAMSTIME

A Stareye Hermit Crab (Dardanus pedunculatus) transports several Tricolor (or Hitchhiking) Anemones (Calliactis tricolor) riding along on its borrowed shell.

Because of the toxins in their nematocysts, anemones are members of an elite club of coral reef organisms with therapeutic potential. The Sun Anemone found on many Caribbean reefs, for example, contains a substance within its venomous tentacles — *stichodactyla* (stik-oh-DAK-tilla) *toxin*, or *ShK* — that led to creating a chemical analogue with promise, potentially a treatment for multiple sclerosis and other auto-immune diseases.

Other than corals, few reef creatures have the capacity to outlive us bipeds. But some anemone species may be capable of becoming centenarians, even multi-centenarians. Young or old, anemones are a declaration of the wondrous nature of coral reefs, lending them color and texture.

FAREWELL TO "GRANNY"

"A famous object of curiosity to the sightseers who visit the Royal Botanic Gardens in Edinburgh has just passed away in the person — if the term is admissible — of the old sea anemone popularly known by the affectionate nickname of 'Granny.' This venerable specimen of the curious class of zoophytes which belong to the very borderland that separates the animal from the vegetable world was certainly 60 years old. . . ."

— *New York Times,* November 2, 1887

Plucked from her seaside home by the Scottish naturalist Sir John Graham Dalyell, 60-year-old Granny was not a native of a tropical coral reef but a cold-water species. "She" was supposedly named for her impressive procreational talents. Her hundreds of kin earned nary a mention in the obituary. But it was noted that Granny could "hardly be reproached with gluttony, since [her] food was simply half a mussel dropped regularly once a fortnight into the membranous aesophagal tube which does duty for a mouth."

A portrait of "Granny," drawn from life. Source: Project Gutenberg Ebook of Glimpses of Ocean Life, or Rock Pools and the Lessons They Teach. *John Harper, 1860, published by T. Nelson & Sons, London.*

3
Aristotle's Sponges

Simple in structure, little different from the first sponges that spread their mats on ancient rocks, they bridge the eons of time.

— Rachel Carson, *The Edge of the Sea,* 1955

In the calm seas off the Greek island of Lesbos, a lean diver surveys the depths. In one hand, he clutches a sharp cutting tool and in the other, a heavy rock. A long line tethers his waist to the boat. Inhaling deeply, he plunges into the sea. The line peels overboard, chases him down. His descent is accelerated by the rock, his progress marked by other divers in the boat. Minutes later, he returns, gulps hungrily at the air, clutching a basket filled with lumpy, unmistakable shapes. The sponges will be dried, then put in sacks for market. The year is 344 BC.

There's another man in the boat, one who peers longingly into the depths, wishing he had the skills to join the diver below. There not to assist but to learn from the divers, the 40-year-old man is insatiably curious about the natural environment. Perhaps the world's first marine biologist, he questions the divers: "Do the sponges sense your arrival on the sea bed? Do they hear you? Will they continue to grow if placed in sea water? Do you think them animal or vegetable?" Their observations later find their way into the man's book, *History of Animals.* It will be one of the most influential natural history books ever written. The author is Aristotle.

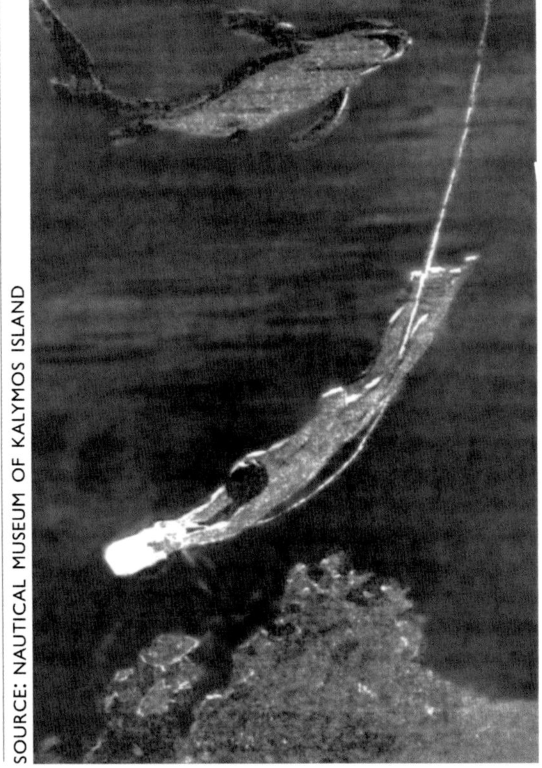

SOURCE: NAUTICAL MUSEUM OF KALYMOS ISLAND

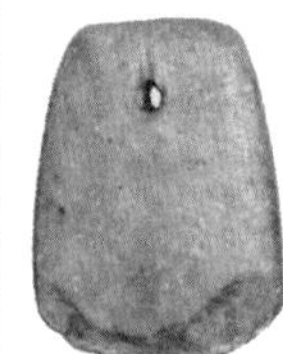

In Aristotle's time and for many years after, Greek sponge divers used a stone called a skandalopetra (scan-de-lo-PET-ra) *to assist in their descent.*

Aristotle spent two years on Lesbos, there to describe its rich marine and terrestrial flora and fauna. The sea sponges he found especially intriguing. They looked like plants, and the divers had told him they were rooted in place. Yet when Aristotle put them in terracotta vessels filled with sea water, they acted very much unplantlike, excreting from multiple pores. When puzzled like this, Aristotle was honest with his future readers:

> *... as regards certain things in the sea, one would be at a loss to know whether they are animals or plants... how to classify them is unclear.*

To see a living sponge is to peer hundreds of millions of years into the earth's past. Fossil records depict little in the way of "spongy" alterations over time; it's as if these simple creatures had hit upon something good and were sticking with it. They'd been knocking about for a long time when the new kids on the block — *Tyrannosaurus rex*, *Triceratops* and their Cretaceous-period cronies — lumbered onto the stage. Stubbornly clinging to a basic design plan, normally a death sentence to creatures ambivalent toward evolution, the sponge is a prime example of survival of the fittest. As 97% of all species that ever existed have died off, sponges have thumbed their *spicules* (glass-like skeletal needles) at Mother Nature's cataclysms.

Back on Lesbos, the divers would have told Aristotle there was little chase involved in pursuing their quarry. Nearly all sponges are sedentary and content to remain in one spot for a lifetime. Obsessive about one thing only, the transport of water, they have no apparent organs, they lack a central nervous system and they show no obvious means of producing little sponges. How can they possibly be animals? Therefore, let's call them plants and be done with it. So the thinking went for 2,000 years.

Then the 19th century arrived, a time when remarkable geologic findings became almost commonplace. Fossils of everything from oysters to mastodons were prompting a re-examination of scientific thought. Spirited

HELMUT CORNELI / ALAMY

Barrel sponges can be some of the oldest residents of a tropical coral reef. Here anthias swim near a Xestospongia testudinaria, a species of barrel sponge at Puerto Galera, Mindoro, Philippines.

debate focused on the age of the earth and how its creatures came to be. Of the numerous riddles at hand, the sponge was one of the most puzzling. One researcher, the Scotsman Robert Grant, wrote "All facts known about the sponge have remained where Aristotle left them." With regard to his study subject, Grant refused to "throw up the sponge" — in boxing parlance, give in to defeat (more common now is "throw in the towel"). He was the first

JEFF MILLER

An Orange Tube Sponge (Aplysina fistularis) seems to be smoking, but the illusion is due to a harmless dye added to illustrate the water circulation through the sponge — in through the pores in the sides, out through the osculum *(OSS-kew-lum), the opening at the top of the central cavity.*

to describe how sponge eggs disperse, propelling themselves about with tiny whiplike structures called *flagella* (flah-JELL-uh).

Because of Grant and others, we know that sponges are the most primitive group of multicellular animals on the planet. They lead a simple life: Pump water in, extract food particles, expel waste, pump water out, produce little sponges.

Where abundant, sponges add a texture and hue to the marine world. Orange, yellow, green, purple, violet and scarlet are just a few of the sponge colors brightly decorating coral reefs and mangroves. From pencil-eraser- to man-sized, sponges come in the shapes of bowls, barrels, tubes, balls, vases, ropes, candles and branching

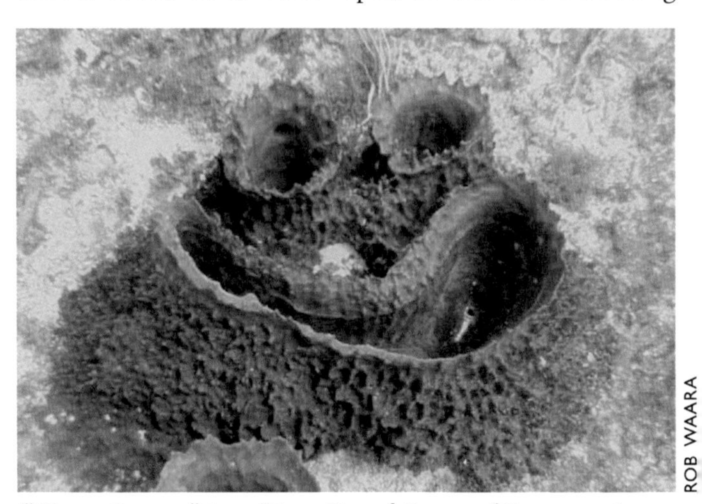

ROB WAARA

"Happy Sponge" is a Giant Barrel Sponge (Xestospongia muta), photographed at a depth of about 27 meters in the Virgin Islands Coral Reef National Monument, off the island of St. John.

horns. Others are less creative in form, encrusting and taking on the shape of whatever it is they overgrow. There are the Stinker Sponge, Touch-me-not Sponge (and you shouldn't, for the nasty sting they impart), Chicken Liver Sponge, Dead Man's Fingers and Bumpy Ball Sponge among the more than 5,000 species. Long-lived, some of these species may be viable for as long as 1,000 years.

The harvest of true bath sponges, a fishery virtually nonexistent now due to overfishing and pollution, stretches back 4,000 years to the Minoan civilization of Crete. Roman soldiers found them useful for carrying drinking water. Aristotle wrote of the sponge of Achilles, so named for its use "as a lining to helmets and greaves (leg armor), for the purpose of deadening the sound of the blow..." He would have been amused to learn that a small group of dolphins in Western Australian waters carry sponges about to protect their sensitive snouts while uprooting tasty tidbits from the seabed. Curiously, the practice seems to be passed only from female dolphins to their female offspring. Decorator crabs (there are several species) hijack sponges, carting them about as camouflage or as a coat of armor.

Still other reef denizens, such as shrimps, various worms and brittle stars (closely related to starfishes, or sea stars) value the sponge as a sound real estate investment and move on in. Animals that live in sponges are quite well protected from being eaten, since many sponges rank far down on the scale of reef snacks, toxic as they

CAROLINE S. ROGERS

Mangroves are important refuges and feeding grounds for many reef animals. This Queen Angelfish (Holacanthus ciliaris) can choose from several sponge species encrusting the prop roots of the mangrove trees in St. John, U. S. Virgin Islands.

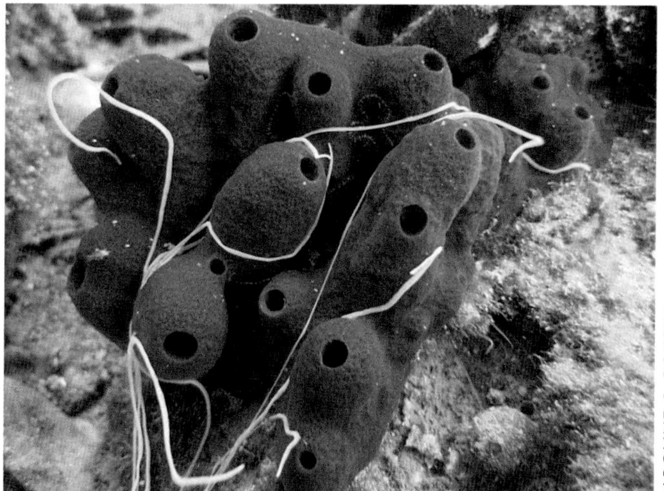

CAROLINE S. ROGERS

Here a red sponge is "fouled" by the tentacles (they can be as much as a meter long) of a Spaghetti Worm (Eupolymnia crasscornis). Also, barely noticeable is a brittle star curled within the recesses of the sponge.

are and laden with silica spicules (essentially glass shards). In the Caribbean, starfishes seem to get around this and appear to find sponges quite tasty, as do Hawksbill Sea Turtles, angelfishes and several species of parrotfishes. Caribbean populations of Hawksbills feed almost entirely on sponges, many of which are poisonous to other vertebrates and contain substantial numbers of spicules. One study found that the digestive tract of an adult Hawksbill could contain almost half a kilogram of these needles!

One might think the delicate balance between competing life forms on coral reefs would bypass such passive creatures as sponges — but nothing could be further from the truth. Sponges need space, and if given the chance they will grow to shade or smother corals and algae, both of which need light for the photosynthesis required for their nutrition.

Aristotle was onto something when he intuited that there was more to the sponge than meets the eye. He would have been astonished at how far-reaching were the secrets held by such a simple-looking structure. An array of medicinal derivatives — anti-fungals, antibiotics, anti-tumor and anti-HIV products — have been discovered thanks to biomedical research on compounds produced by sea sponges, research that started at the edge of the Adriatic when a Greek naturalist beheld a sponge and began to take notes.

LIVING INSIDE A SPONGE

A sponge is a somewhat unusual address for snapping shrimps, noisy little creatures, maxing out at about 5 cm long. Wielding a prodigious claw used to stun prey or for defense, snapping shrimps, also called pistol shrimps, produce many of the characteristic snapping and popping sounds often heard on reefs. A shrimp that moves into a Loggerhead Sponge (*Spheciospongia vesparium*) at a youthful age and discovers the merits of a "gated" community within the sponge, may then become imprisoned by its own size as it grows. We can imagine it wandering the tubed mazes as if trapped in a chamber of horrors. Sometimes an entire colony of shrimp will settle into a sponge, perhaps a sponge species with easier in-and-out access.

WATERFRAME / ALAMY

A pair of snapping shrimps living inside a vase sponge near the island of Sulawesi, Indonesia. Some shrimps living in sponges are akin to social insects like bees and termites, and are the only known social species in the marine environment.

4

Tale of the Tunicate's Tail

You've asked me what the lobster is weaving there with his golden feet?

I reply, the ocean knows this.

You say, what is the ascidia waiting for in its transparent bell? What is it waiting for?

I tell you it is waiting for time, like you.

— from *The Enigmas* by Pablo Neruda

In 1494, when Christopher Columbus "discovered" Jamaica on his second voyage to the New World, he sailed along the north coast of the Caribbean island in search of fresh water, anchoring in what appeared to be a promising site. His search for water proved fruitless and the disappointed navigator moved on. But let's imagine that he was distracted by the bizarre little blue sacs among the other creatures his anchor dredged up from their attachment to the sea bed. Perhaps he examined one and noted its vaguely translucent appearance, unusual for what he likely assumed was an underwater plant. Maybe he was even startled when, without warning, the tube squirted a jet of water in his direction wholly disproportionate to its size.

CAROLINE S. ROGERS

A colony of Bluebell Tunicates (Clavelina puertosecensis) on a green sea plume (a type of soft coral) in the Caribbean Sea

Clavelina puertosecensis, commonly known as the Bluebell Tunicate or Blue Sea Squirt, is one of more than 2,000 species found worldwide and at all depths. Tunicates vary in size from about 0.5 cm up to 15 cm long. They are among the most common invertebrates found on coral reefs, but are often overlooked or mistakenly identified as sponges. The name "tunicate" comes from the cellulose-like covering that all of them possess and that provides a protective "overcoat," or tunic. While some are free-swimming, those we are most likely to come into contact with, sedentary and baglike, belong to a class known as Ascidiacea, (uh-SID-ee-ACE-eeya) from the Latin term for wineskin.

Nearly 500 years after Columbus encountered the Bluebells on his anchor, the same area witnessed another discovery, this time underwater. Marine biologists recognized the blue organisms for what they really are — a species of tunicates — also known as *sea squirts,* a name Columbus might have appreciated if our little fantasy had taken place. In 1974, the Irish marine biologist and world authority on sea squirts Ivan Goodbody gave it the species name *puertosecensis.* "When Christopher Columbus discovered Jamaica," he wrote, "he sailed along the north coast looking for fresh water. Disappointed in what we call Discovery Bay, he named it Puerto Seco, Dry Harbourland... and that is where we got the name for the ascidian."

In its adult form, the Bluebell, like other ascidians, colonial or solitary, lives attached to the sea bed, rocks, pilings or even to the branches of corals. The larval tunicate, however, gives no hint as to what it will eventually become; it looks more like a tadpole.

The tunicate larvae swim about for a few days before anchoring themselves as they begin what counts as one of the most thrilling transformations in the natural world. Once attached, the tadpole signs on for a total makeover, morphing from a free-swimming larva complete with brain, primitive eyes and a tail, into the permanently attached spongelike sack it will forever be. The transformation complete, the tunicate has become a highly efficient filter-feeding organism, quite content to spend its days moving water into one orifice and out of another.

SERGIO LLAGUNO / DREAMSTIME

The solitary Gold-mouth Sea Squirt (Polycarpa aurata)

But a simple sac it is not. These uncomplicated-looking animals have elaborate nervous, digestive and reproductive systems. Indeed, their circulatory systems would drive a cardiologist mad, with a heart that periodically completely reverses the direction of its circulation.

Long before Dr. Goodbody began his Jamaican investigations, other biologists were drawn to the ascidians. One, a young Russian embryologist named Alexander Kowalevsky, was fascinated by the ascidians he collected from Mediterranean waters. Kowalevsky proved that tunicates were much more complex than previously thought. In 1867, the year the United States purchased Alaska from his native land, Kowalevsky made a startling discovery. In the larval tunicate's tail, he found a "support rod," a *notochord* (NO-tow-cord), precursor of that which keeps us all from keeling over into our soup — the backbone. He suggested that tunicates were the ancestors of ancient fishes, a bold assertion that eventually proved accurate, parachuting ascidians right into the same phylum — Chordata (kor-DAH-ta) — as humans and all other species possessing central nerve cords, whether encased in a vertebral column or not.

We now know that about 365 million years ago, some fishes, not content with the life aquatic, wriggled onto land. This would be the biggest leap in vertebrate evolution so far, an elegant connection linking tunicates to gill-breathing fishes to pioneering marine scientists, Nobel Prize–winning poets such as Pablo Neruda — and the rest of us.

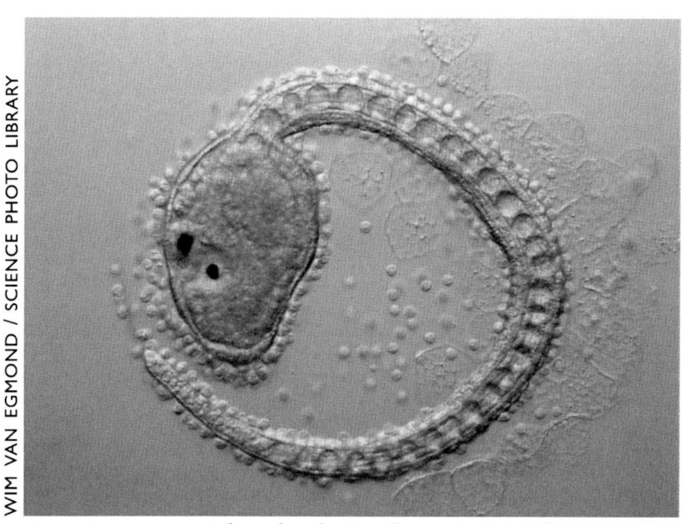
WIM VAN EGMOND / SCIENCE PHOTO LIBRARY

A tunicate egg with a developing larva inside, showing the notochord that establishes the relationship between tunicates and all other animals that have a central nerve cord.

FROM REEF TO PHARMA

Intrigued by all things related to marine biology, Ivan Goodbody harvested the Mangrove Tunicates he found offshore of his adopted country, Jamaica. He sent them off to European scientists, who would ultimately use them to develop trabectedin (truh-BECK-tuh-din), or Yondelis® (yon-DELL-iss) a medication useful for treating various soft-tissue cancers.

CAROLINE S. ROGERS

The Mangrove Tunicate (Ecteinascidia turbinata) grows in dense clusters.

5
Sea Slug Fest

In Greek mythology Glaucus (GLOCK-us) was a fisherman rescued by the sea gods Oceanus and Tethys, who then added him to their lofty ranks. All was going swimmingly for the new merman until the day he cast his eye on the lovely mortal Scylla (SILL-uh). In a rage of jealousy, the goddess Circe turned Scylla into a six-headed monster. Justifiably upset with her new look, Scylla took to biting off the heads of passing sailors.

ANTONIO TEMPESTA, 1606

Scylla and Glaucus*, an illustration for Ovid's narrative poem* Metamorphosis

The tragedy plays itself out in the real-world lives of two *nudibranchs* (NEW-dih-branks): *Scyllaea pelagica* and *Glaucus atlanticus.* Nudibranchs, often called *sea slugs,* are a type of marine gastropod — a marine snail minus the shell. *Scyllaea* (*SILL-ee-ya*) passes its days floating in *Sargassum* seaweed, though it might also be seen on other kinds of algae or even on mooring lines. Perfectly camouflaged, it dines not on sailors but on errant morsels of jellyfish.

NORBERT WU / MINDEN PICTURES

Scyllaea pelagica provides an exception to the rule among nudibranchs. Instead of being brightly colored, it matches its environment almost perfectly.

S.ROHRLACH / ISTOCKPHOTO

Glaucus atlanticus, the Blue Dragon, is a small sea slug found in the open ocean, a relative of the many nudibranchs associated with coral reefs.

Glaucus is a stunning pelagic species, hanging out in the open sea, occasionally washing up on beaches. Upset perhaps over the loss of his one true love, *Glaucus* sometimes teams up with fellow species members, and like a pack of hungry wolves, they surround and devour Portuguese Man-o'-War "jellyfish" down to the last bite.

Glaucus and *Scyllaea* are just two of approximately 3,000 sea slug species, many of which can be found gliding, sometimes swimming, around coral reefs. The name "sea slug" is misleading, for nudibranchs are not at all like those slimy nibblers that visit your garden. Sea slugs are flamboyant, flaunting frilly outfits and sporting whimsical accoutrements, body bling without equal in the marine world. They look like the creation of an artist willing to push the bounds of extravagance with their undulating flaps and sensory organs and fluttering clusters of "fingers." One species is streaked with jet black stripes and fringed in turquoise and gold. There are mottled maroon, lime green, orange and polka-dotted nudibranchs, some that are translucent white, and countless species with polychromatic patterns reminiscent of a Tiffany lamp.

Nudibranchs are named for their "naked gill" structures, called *cerata* (seh-RAH-ta; from the ancient Greek word for horns). Flattened, ruffled or shaped like delicate lace, in most species cerata are on the outside surface of the body.

In many species a pair of sensory organs called *rhinophores* (RYE-nuh-forz) project from the head and are used to smell or taste the water. Rhinophores help guide the sight-challenged nudibranch to its next meal or to a potential mate. For no purpose obvious to us, several members of a species will occasionally tailgate one another, sniffing out "slime trails" in a bizarre exhibition of follow-the-leader.

CIGDEM SEAN COOPER / DREAMSTIME

Some nudibranchs, like this Fabellina species perched on soft coral in the Red Sea, sport cerata distributed all over the top surface of the animal.

HANS GERT BROEDER / DREAMSTIME

The cerata of some nudibranchs, like those of the Anna's Magnificent Slug (Chromodoris annae) shown here, are clustered together in a bunch.

Associated not only with coral reefs, nudibranchs are found in most marine habitats, from intertidal pools down to 1,000 meters. Most nudibranch species are found in the Indo-Pacific biogeographic region.

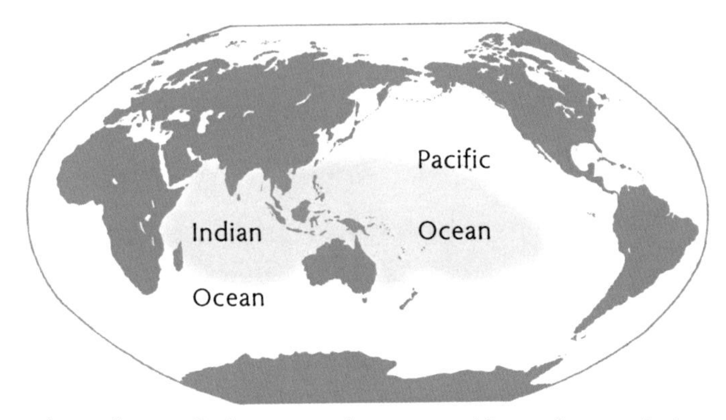

The Indo-Pacific biogeographic region (shown here in light blue) is home to many species of nudibranchs.

Many nudibranchs are not much larger than a pencil eraser. However, the Spanish Dancer won't test your eyesight, coming in at a prodigious 60 centimeters long. Flashing a ruffled scarlet "skirt," its aquatic flamenco dance is not soon forgotten.

Many sea slugs work the day shift, others the night, but almost all spend their adult life on the seabed, gliding on clear mucus laid down by the sole of a flat, muscular foot. Some live unseen lives under rocks or sand, while others are found wedged into sponges or tucked into the latticework of a coral structure. A sharp eye might observe them in seagrass beds or crawling over the rough terrain of hard corals. Occasionally one might be spied slithering across the latticework of a *sea fan* (a type of soft coral). During its wanderings a nudibranch might happen upon a lie-in-wait predator such as a stonefish or frogfish. Lacking any inclination to reverse, the nudibranch just

WATERFRAME / ALAMY

The brilliant coloration of the Spanish Dancer (Hexabranchus sanguineus) warns predators that it tastes awful.

keeps on trucking, carefree or careless, one can't be sure. Perhaps through indifference, or more likely unwilling to give away its location, the fish suffers the indignity of having a slime secreter crawl over its snout, waiting patiently for the colorful interloper to be on its way.

How do they get away with it, these quirky life forms that look like the perfect meal for many a sea creature? In a habitat where most small animals go to great pains not to be noticed, to what end all that color? And how do some species manage to go for a cheeky little swim in the wide blue expanse without being picked off by a hungry mouth attached to a streaking body? For the answers, we jump back some 200 million years to the time when nudibranchs began testing alternative defensive systems absent the protective shell of their snail relatives.

DAVID FLEETHAM / ALAMY

A Crested Nembrotha nudibranch (Nembrotha cristata) makes its way across a patch of tunicates in the Philippines.

One of those strategies settled on what has become a sophisticated arsenal of weaponry, an arms cache respected by most predators. A number of nudibranchs dine on such things as corals, fire corals and anemones. As you may remember (from pages 8 and 9) all of these "food" species are in the phylum Cnidaria, creatures that possess nematocysts, specialized structures with tightly coiled fibers tipped with toxic harpoons. When triggered by a too inquisitive predator (or as some of us have learned, an errant bit of human anatomy), the capsules discharge and the venom is injected. But nudibranchs are immune to these stinging cells. They gobble up cnidarians and sequester the nematocysts in the tips of their cerata. Think of the nudibranch as a medieval castle, the nematocysts a battalion of mercenaries, their poisoned arrows-for-hire ready to protect their host at the slightest nibble.

Furthermore, some nudibranchs feed on toxic sponges with no ill effect, "borrowing" the poisons and storing them in their tissues. It seems most predators need to encounter the arrows, or taste the poisons, only once to learn to steer clear for the rest of their lives.

Nudibranchs that are toxic and "armed to the gills" have no real need to concern themselves with camouflage. Quite the opposite in fact. The effect of all those showy colors and designs, known as *warning coloration,* or *aposematic* (AP-po-se-MAT-ic) *coloration,* seems a way to announce "Here I am, dine at your own risk. You won't forget it!"

Oh, and they do have one last trick up their cerata. If all else fails, a nudibranch can "launch" itself from the bottom, writhing its way to safety.

In nature, there's always somebody who figures out a way to level the playing field. Some crabs, for instance, have learned to pluck the cerata from a nudibranch, ridding their catch of the harmful nematocysts before eating it. Sea stars, somehow immune, will dine on nudibranchs if given the chance. And, presumably after a bit of trial and error, some wrasses and pufferfishes discovered that the strength of a nudibranch's chemical discharge can be nullified by repeatedly sucking the creature in with a gulp of water, then spitting it out, "washing" and rewashing it until it's safe to eat.

Of course sea slugs need to eat as well, and they are carnivorous. Besides cnidarians, they prey on colonies of tiny creatures called *bryozoans* (bry-uh-ZO-enz) or moss animals, as well as sponges. Some even supplement their energy intake via a process called *kleptoplasty*: The Lettuce Sea Slug, not a nudibranch but a close relative, feeds on green algae. Some of the algal cells are stored in the sea slug's fleshy appendages, where photosynthesis continues to take place, providing energy-giving sugars for the zooxanthellae's new host.

Last but not least, some sea slug species aren't averse to making a meal of each other. Given the opportunity (and presumably a hearty appetite) a hungry nudibranch will even begin munching its partner while *in flagrante delicto*.

Speaking of which, all nudibranchs can be described by the gloriously descriptive phrase *functional simultaneous*

CAROLINE S. ROGERS

Elysia crispata, the "solar-powered" Lettuce Sea Slug, uses the chloroplasts it eats to convert sunlight into nutrients.

F. SCHNFIDER / ARCO IMAGES GMBH / ALAMY

A Dusky Nembrotha nudibranch (Nembrotha kubaryana) cannibalizes a member of its own species.

hermaphrodites. It's a fancy way of saying that each individual has the necessary plumbing to both produce viable eggs and fertilize them. Because their reproductive pores are located on each animal's right side, nudibranchs line up starboard to starboard, head to tail. Both individuals act as males, swapping sperm packets, the penis of each inserted into the female duct of the other. Some species then cast their eggs into the water column and hope for the best, but most lay them on an appropriate surface and call it a day — or a night, as the case may be — leaving the eggs behind as they crawl away. Once hatched, nudibranchs haven't much time to waste; few barely make it past the ripe old age of 12 months.

THE NAPOLEAN CONNECTION

Believe it or not, our knowledge of nudibranchs can be traced to Napoleon Bonaparte. Driven by his thirst for scientific knowledge — as well as his appetite for military adventuring — Napoleon mounted what came to be known as the Expedition d'Égypte in 1798. In addition to thousands of soldiers, the massive fleet included 167 scientists whose goal was to study the regional flora and fauna. Among them was a precocious 21-year-old naturalist, Jules-César Savigny. Savigny was fascinated by marine creatures, and his expedition work resulted in beautifully engraved plates of marine organisms from the Red Sea, including nudibranchs. The plates found their way into a classic reference, the *Description de l'Egypte*, a comprehensive study of the region's cultural and natural history spanning a 17-year period of research. Though Savigny went blind at an early age and depended on others to complete his research, he, and indirectly Napoleon, are due the credit for providing us with one of the earliest representations of these fascinating creatures.

Jules-César Savigny, memorialized in a portrait by André Dutertre from Louis Reybaud's Histoire Scientifique et Militaire de L'expédition Française en Égypte

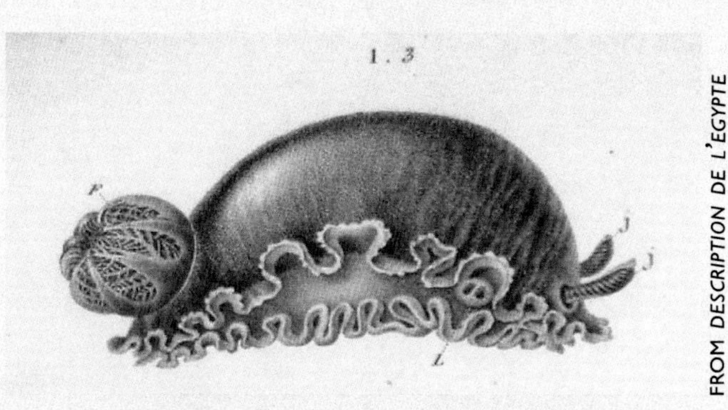

FROM *DESCRIPTION DE L'EGYPTE*

Savigny was one of the scientists who accompanied Napoleon Bonaparte on his Egyptian expedition. The marine flora and fauna he found, such as this nudibranch, were documented in engravings.

6

Beguiling Blennies

Blennies are relatively small, difficult to identify and well off the radar screen for many divers and snorkelers. But if one takes the time to carefully look for and observe these little tadpole-shaped fishes, their charisma and visual appeal are ample reward. The color patterns of blennies, their nesting activity and their threat and courtship displays are enough to engage a marine naturalist's curiosity for hours on end. Not to mention their curiously appealing tendency to look in one moment as if all the world is a source of wonder and in the next act as if all the world is out to get them.

Even Aristotle was drawn into their universe. On the island of Lesbos in the northern reaches of the Aegean Sea, he perched for hours on a tiny platform over a lagoon comparing the habits of a Giant Goby (*Gobius cobitis*) with those of the Intertidal Blenny (*Parablennius sanguinolentus*). Keep in mind, this was some 2,400 years ago and Aristotle was grappling with fundamental problems that included form and function as it related to these two similar looking, yet very distinct creatures from different families. In cupping them in his hand, Aristotle noted the differences in their fins — the goby's sucker-shaped, the blenny's like rays. Yet both sets of fins did much the same thing, anchoring the fish to the bottom during turbulent sea conditions.

FISH FIN BASICS

When you read about fishes, it's helpful to know some basic terminology for one of their most prominent features — their fins. Shown here is a generic fish body with the five most commonly encountered fin types, typically used to propel, stabilize or steer a fish as it swims.

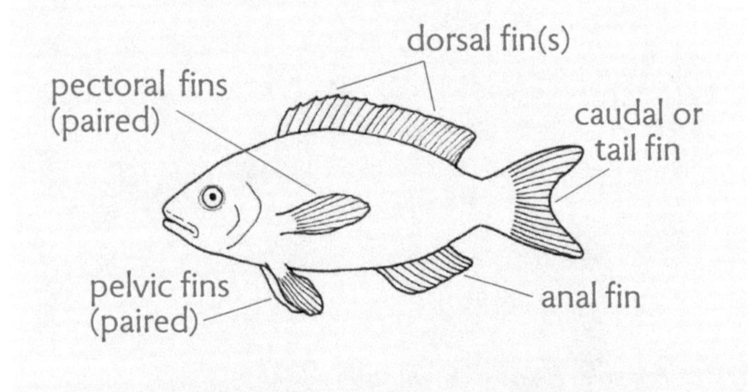

SUSAN HELLER

We can only imagine what joy Aristotle might have found in studying the wealth of blennies known today. He would be astounded, for example, to learn of the sheer number of species, some 900 and counting; among them are con artists, a terrestrial gadabout and dozens of what look for all the world like runway models, complete with jaw-dropping colors and decorative appendages — aptly referred to as *ornaments.* Such additions include elaborate crests and head *cirri* (SEAR-ee), which are tentacle- or tassel-like protuberances.

GARY BELL

Like other blennies, the charismatic Segmented Blenny (Salarias segmentatus) uses its pelvic fins to brace itself.

JONATHAN LAVAN

The Red-spotted Blenny (Blenniella chrysospilos) is ornamented with head cirri.

Blennies may also have unusually large dorsal fins, whose primary function seems to be to attract a mate, or a photographer's lens. In *The Descent of Man, and Selection in Relation to Sex* (1871) Charles Darwin hypothesized that such traits were sexually selected, noting that "with some blennies... a crest is developed on the head of the male only during the breeding season.... There can be little doubt that this crest serves as a temporary sexual ornament, for the female does not exhibit a trace of it."

JONATHAN CHURCHILL / SHUTTERSTOCK

The Sailfin Blenny (Emblemaria pandionis) (shown at the left standing on his tail in the sand) displays a disproportionately large dorsal fin.

Male blennies establish territories around a gap or hole in the reef, where they pin their hopes on nesting success. Their aggressive displays toward other males and their courtship maneuvers toward females include quivering, head nodding and dorsal-fin flashing. An especially creative and agile blenny might launch itself into the water column and run through an acrobatic repertoire that can include circle, figure-8 and zigzag swims. The female is attracted into the residence, where if she's satisfied with the male's digs, she'll deposit her eggs — and then she leaves. After he fertilizes the eggs, the stay-at-home father guards the nest from predators and conducts basic housekeeping activities until the eggs hatch.

Blennies have an impressive variety of dentition and jaw types. Many species have rows of short canine-like teeth, while others have minute combs arranged in rows, useful for grazing on algae and other detritus on the reef. Still others have large curved canines used for fighting as well as defense against predators. Blennies feed on plankton, small invertebrates and the skin and scales of other fishes. These are small fishes, rarely exceeding forefinger length, and they spend most of their time perched on the seabed, where they are easily approached — up to a point. They are found around the world, from tropical coastlines to the Antarctic Peninsula, and in a multitude of habitats, including tropical coral reefs and mangroves. They frequent seagrass beds and temperate kelp habitats and are found in fresh, salty, and brackish water. They hang out in tidepools, tuck in under rocks and boulders and burrow into sandy bottoms. The upshot is that watchful reef visitors, be they snorkelers or scuba divers, stand a fair chance of finding blennies regardless of the habitat, though blennies do have the ability to change color to blend in with their surroundings.

Humans often confuse blennies with gobies. To us the two kinds of fishes look and behave alike, but a good rule of thumb is that blennies typically perch with the body curved to one side in "casual" repose, while gobies are more stiffly aligned in straight positions.

The lack of a swim bladder prevents most blennies from cruising the water column. Without this organ they can't adjust their buoyancy, and so they naturally settle

NED DELOACH

A pair of Molly Miller Blennies (Scartella cristata) in a nesting place found and defended by the male (right).

CAROLINE S. ROGERS

The Hairy Blenny (Labrisomus nuchipinnis) in repose, its body curved

to the bottom. On coral reefs many take up residence on brain or boulder corals, just their heads peering out of tiny holes, tracking the world with googly eyes that don't miss a trick — or a snack from the reef's ongoing buffet of floating organic tidbits. Others prefer to hang out in

DAMSEA / SHUTTERSTOCK

A Spotjaw Blenny (Acanthemblemaria rivasi) shelters in a Brain Coral near Bocas del Toro, Panama.

DURDEN IMAGES / SHUTTERSTOCK

Just about 4 cm long, this Diamond Blenny (Malacoctenus boehlkei) uses its pelvic fins as a sort of tripod as it hides among the protective stinging tentacles of the Giant Caribbean Sea Anenome (Condylactis gigantea).

coral rubble, on sea fans, or on blades of sea grasses. An unoccupied shell comprises a dandy *pied à mer* for the enterprising blenny. And after an initial adjustment during which they put up with a sting or two, the Diamond Blenny and others move safely in among the protective tentacles of sea anemones.

The Bluestriped Fangblenny also hides in the reef, but in an entirely different way, hanging out at a reef "cleaning station." These are often large coral heads where fishes of all persuasions line up as if at a car wash, patiently waiting to have mucus and ectoparasites removed by gleaners such as the juvenile Bluestreak Cleaner Wrasse. At times this fish wash may be visited by close to 200 clients per hour!

A patient fish watcher might spot what seems to be another wrasse waiting nearby to service its customers. It looks for all the world like the Bluestreak, with the same colors and patterning. The customer approaches, fully expecting a nice grooming. Instead, the attendant turns out to be an imposter, darting up to the unsuspecting victim for a quick bite of fin or scale, providing the rare view of a fish jumping in surprise. The conniver is the Bluestriped Fangblenny, one of the world's best examples of *aggressive mimicry*: One species, in this case the Fangblenny, frequents the same area and mimics the color and appearance of its so-called model, here the Bluestreak Cleaner Wrasse. The mimic obtains a meal from the unsuspecting prey and is unlikely to be attacked because of its resemblance to the model. (In another predatory strategy, the sneaky little fangblenny can even alter its color and striping to conceal itself within a fish shoal, or school, from which it launches attacks on passing fishes.)

Fangblennies, also called sabertooth blennies, belong to a group of more than 50 species that come heavily armed and swim fearlessly through open water, quite the opposite of their unarmed relations that wouldn't dare risk leaving the security of their blenny burrows. Those fangs also come with an additional fright: the ability to inject venom via a groove in their curved canines.

The fangblenny was introduced to science in 1852 by Pieter Bleeker, a Dutchman who rose to become one of the world's greatest ichthyologists, one whose influence has lasted through today. Bleeker came from a humble background to become a physician with as keen an interest in zoology as in doctoring. Joining the army as a medical officer, off he went to Batavia (present day Jakarta), where the 23-year-old must have pinched himself many times over for his good fortune. At the time, the Indo-Pacific was a hotbed of marine biodiversity, largely unknown and just waiting for the right person — Bleeker — to unveil its mysteries. By the end of his 18-year stint in Batavia, he had collected and described over 12,000 fishes, most of which were shipped to the Rijksmuseum van

NED DELOACH

The Bluestriped Fangblenny (Plagiotremus rhinorhynchos) (inset) is an aggressive mimic of the juvenile Bluestreak Cleaner Wrasse. The blenny sometimes lies in wait near the wrasse's cleaning station to attack unsuspecting customers.

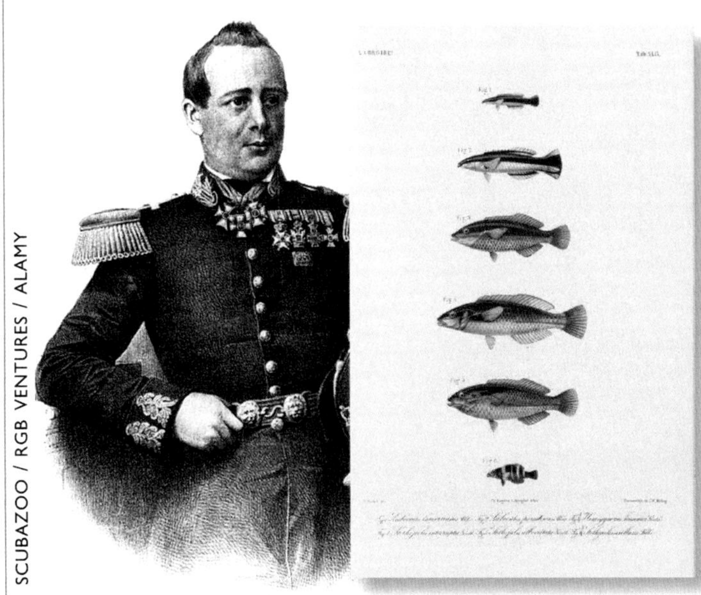

SCUBAZOO / RGB VENTURES / ALAMY

A portrait (artist unknown) of Dr. Pieter Bleeker along with plate 54 from Bleeker's Atlas Ichthyologique. *The two fishes at the top of the page are the Bluestreak Cleaner Wrasse, recognized today as Labroides dimidiatus.*

Natuurlijke Histoire (RMNH) in Leiden, Netherlands, where scientists still use them as study specimens.

Bleeker's contacts throughout the archipelago were extensive, and they often sent study specimens from government postings that he himself had little hope of visiting. He also bought specimens from fishermen and was fond of roaming Pasar Ikan, the local fish market, where we can picture vendors calling out to the doctor they knew would be interested in a recently arrived strange-looking species. Imagine his excitement loading a yet to be described grouper, for instance, onto a horse-drawn carriage for the trip back to his dissection lab.

Bleeker published 500 papers, most on fishes of the Indo-Pacific. He described a phenomenal 1,925 new species, more than any other ichthyologist, 40% of which are still valid — that is, still considered accurate by ichthyologists today. Upon his return to the Netherlands in 1860, he began work on what would become his *magnum opus*, the *Atlas Ichthyologique*, a compendium that summarized much of his Indonesian labors and a book still highly regarded today for the accuracy of its more than 1,500 illustrations.

We can only surmise how he came by the diminutive fangblenny, about as far as one can conjure from the typical reef fish that would have been hawked by fish mongers. Perhaps as Bleeker dissected our imagined grouper, much to his surprise the body of the hapless blenny tumbled onto the table, demonstrating that the aggressive mimic doesn't always win, and proving that a version of that old saw, variously attributed to Abraham Lincoln, P. T. Barnum and others, even prevails on the coral reef: "You can fool all the fishes some of the time and some of the fishes all the time, but you cannot fool all the fishes all the time."

BLENNIES ON LAND

The strangest blenny of all is a species found in the islands of the Western and Southern Pacific. The Pacific Leaping Blenny, also called a Rockskipper, has successfully transitioned from an underwater to an intertidal existence. With a quick twist and flick of the tail it moves effectively, if not gracefully, about rocky areas. As long as it keeps moist, the Leaping Blenny can breathe through both its skin and gills, providing evolutionary biologists with tantalizing evidence of how animal life may have come to colonize the land.

J'NEL / SHUTTERSTOCK

The Pacific Leaping Blenny (Alticus arnoldorum) is a comb-toothed blenny that can live out of the water on exposed rocks and reefs for hours at a time, as long as it doesn't dry out.

7
Winged Flyers

Stingrays are highly adaptable, primarily warm-water fishes. They range from the size of one's hand to the Short-tail Stingray, weighing in at a prodigious 350 kg or so with a wingspan of more than 2 meters. They cruise about sandy bottoms, including those around coral reefs, often burying themselves up to the eyeballs to hide from sharks, their only predator other than humans. To see one gliding through the water as if by magic can be thrilling, their appearance somehow other-worldly in a place heavily skewed toward finned forms.

The exoticism of the stingray has not gone unnoticed. The name itself is freighted with a certain cachet: There's the Corvette Stingray sports car; a "Stingray" phone-tracking device is used by law enforcement; and the iconic silhouette pops up on multiple internet sites. Stingray leather products include cell phone covers, shoes, motorcycle seats, even Samurai sword handle wrap.

And then there's Stingray Point, a curious name hitched to a midsummer's day in 1608. Far from the nearest coral reef, the site lies on a smidgen of land washed by the lower Chesapeake Bay. While probing this new frontier, the English explorer Captain John Smith found his ship grounded between tides. Making the best of the temporary setback, he and his men disembarked for a bit

CAROLINE S. ROGERS

A school of little Atlantic Silverside (Menidia menidia) frames a Southern Stingray (Dasyatis americana) in shallow waters near St. John, Virgin Islands National Park.

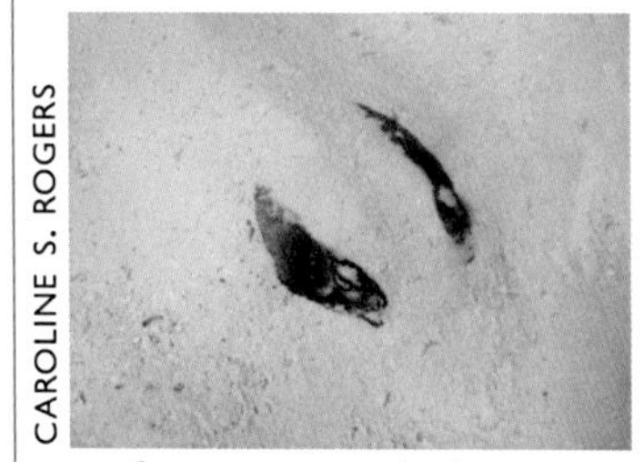

CAROLINE S. ROGERS

Left: A stingray hiding in the sandy bottom of St. John's Leinster Bay, revealing only its distinctive eyes and behind them its spiracles. Right: A Chevy Corvette Stingray borrows that look.

of fishing with, of all things, a frying pan. Finding it "a bad instrument to catch fish with" they took the logical leap, and changed to swords.

In Smith's journal account of the incident, told in third person as was common to the day, the origin of the name "Stingray Point" becomes apparent:

> *But it chanced our captain, taking a fish from his sword (not knowing her condition), being much of the fashion of a thornback [a type of ray], but a long tail like a riding rod — whereon the midst is a most poisoned sting of two or three inches long — bearded like a saw on each side, which she struck into the wrist of his arm near an inch and a half. . . . the torment was instantly so extreme that in four hours had so swollen his hand, arm, and shoulder we all with much sorrow concluded [expected] his funeral, and prepared his grave in an island by, as himself directed.**

STEPHEN FRINK / ALAMY

A barb on the tail of a Southern Stingray

*From pages 59–60 of the *Journals* (John M. Thompson, editor); see page 114 for the full reference.

The grave was never occupied and Smith exacted his retribution by later eating the stingray!

More than 100 million years before that dinner, stingrays arrived in the marine world, fairly late to a party that had been humming along for quite some time. They are members of the class Chondrichthyes (kon-DRIK-theez), a group that includes other kinds of rays as well as skates and sharks. Card-carrying members share several things in common, including: the absence of lungs or swim bladders, multiple gill openings and *cartilaginous* skeletons (made of cartilage rather than bone). But only stingrays carry a long, venomous spine or two in the tail, used exclusively for defense.

The venom has a deleterious effect on the cardiovascular system; even so, most encounters are not fatal. And an encounter may be avoided by slowly shuffling across the sandy bottom to warn any nearby rays so they can get out of the way. Unfortunately, this was not the case with naturalist, conservationist and television personality Steve Irwin, who, when approaching a stingray from behind while snorkeling off Australia's Great Barrier Reef in 2006, succumbed when the fish responded defensively, as if it was being followed by a shark, its barb piercing Irwin's heart.

The stingray doesn't look like a "typical" fish, like the one in the diagram on page 26. The pectoral fins are fused to the fish's head, lending it that disclike Starship *Enterprise* look. And sorting out where such things as gills and mouth are found on the stingray is a little confusing, since they are located underneath. Water is funneled to the gills through openings called *spiracles* (SPEER-uh-klz) on the top side, behind the eyes.

Both sharks and stingrays have an advanced electroreceptive system, a fancy way of saying they can sense minute electrical fields emitted by creatures they like to eat. When the stingray excavates a mollusk from its sandy lair or when a shark homes in on a wounded creature, it relies upon an intricate network of sensory organs called *ampullae of Lorenzini* (AM-pyoo-lee *of* lor-en-ZEE-nee) to detect the electrical fields created by prey species.

The name of these sensory organs comes from a 17th-century Italian physician who specialized in human anatomy. In his spare time Stefano Lorenzini dissected sharks and rays, producing some of the earliest anatomical studies of those creatures. He discovered tubes filled with a gelatinous substance that were connected to pores on the ray's head, each "fastened to a small globe, about the bigness of a coriander-seed...." Lorenzini wondered "whether they were not designed for some... secret use, since the works of nature are never made by chance, nor are her beings multiplied without some necessity...." Lorenzini's analysis was spot on, and without a doubt he would have enjoyed the sight of a stingray hydraulically mining the sea bed, using the sensory organs he had first described in 1678.

In the shared history of humans and stingrays, many threads link the stingray to diverse cultures. In Greek myth, Odysseus was slain by his son Telegonus with a spear tipped with the spine from a stingray. The Roman author, naturalist, philosopher and military commander

ANATOMY OF A STINGRAY

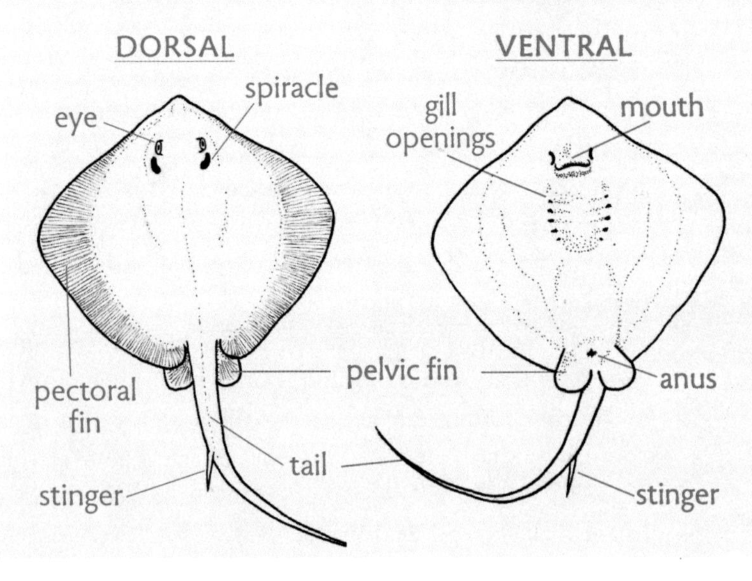

The external anatomy of a stingray is shown here with the dorsal (topside) on the left, and the ventral (underside) on the right. The large pectoral fins are used as "wings" as the fish "flies" through the water. When swimming, the stingray draws water into its mouth; from there the water passes over the gills and exits through the gill openings. When the stingray buries itself in the sand, it exposes only its eyes and spiracles (as shown on page 32), so water for the gills is drawn in through the spiracles rather than the mouth, and passes over the gills from there. For defense, stingrays have one or more stingers on the middle part of the tail.

SUSAN HELLER

Pliny (PLINN-ee)the Elder weighs in with his classically skewed certitude in his epic tome *Naturalis Historia,* or *Natural History*, saying the ray's spine is capable of killing a tree, though we are denied the privilege of learning just how this might occur. The Mayans, after wisely removing the venomous sheath from the stinger, used the spine as part of a penile bloodletting ceremony to appease the gods.

The stingray is also present in some ancestral aboriginal cultures in Australia. For the Australian aboriginal Yolngu group, the stingray, Gawangalkmirri, is a powerful symbol. Yolngu believe the clouds of silt stirred up by nesting stingrays affect the formation of rain-giving thunderheads.

Gawangalkmirri also represents laudable human values for the Yolngu: Stingrays are gentle and social, and they provide a good home for their young. Like the Yolngu, stingrays use their "spears" when threatened, as Captain Smith and many wading beachgoers have unfortunately discovered over the years.

CAROLINE S. ROGERS

This Southern Stingray makes a complete mess of things as it "hunts," no doubt breathing through its spiracles. Meanwhile, a Gray Snapper swimming above it looks to pick off a free meal, a strategy known as kleptoparasitism.

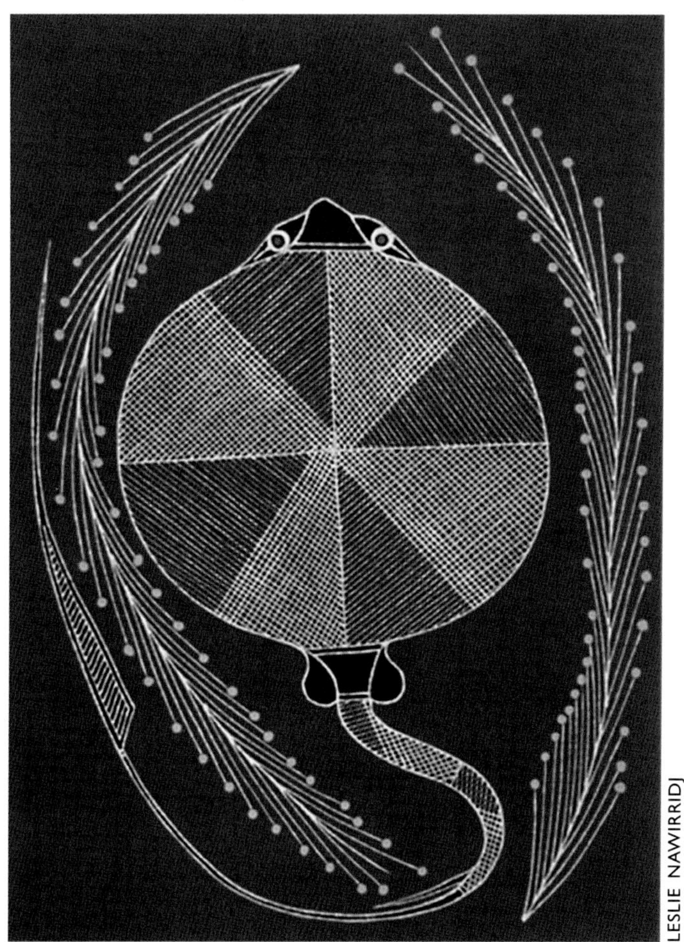

LESLIE NAWIRRIDJ

Symbolic among aboriginal Australians, stingrays are also valued as a food source. Stingray with Seagrass *was painted in the "x-ray" style of the Kunwinjku in West Arnhem Land.*

8

Hina's Eel

*The first to develop a special pond for the moray was Gaius Hirrius, who supplied six thousand morays for Caesar's banquets.**

— Pliny the Elder, *Naturalis Historia,* 77 AD

While finning along a reef edge or scuba diving among the residents of a coral grotto, you might come upon a sight at once both captivating and menacing. Protruding from a hole in the reef or from beneath a ledge, a conical head shows itself, kitted out with impressive teeth. Even more startling is the creature's penchant for opening and closing a very muscular set of jaws. But fear not, diver, the moray eel, with its snakelike shape, has no culinary interest in you, and what appears to be aggressive posturing is simply what the eel must do to pump oxygenated water past its gills.

There are some 200 species of moray eels, from the pencil-sized Ribbon Moray to the 3-meter Giant Moray. They come in a multitude of hues — brown, green,

* The eels Pliny wrote about here might have instead been freshwater lampreys, which the Romans were also fond of, as were the English centuries later.

BRIAN LASENBY / SHUTTERSTOCK

A Green Moray (Gymnothorax funebris) looks out from its sponge-encrusted cave near Cozumel, Mexico.

MORAY EEL FINS

SUSAN HELLER

An eel's fin plan deviates from that of the generic fish body (shown on page 26). For one thing, the dorsal and anal fins are continuous with the caudal fin, if there is one, to form a single wraparound fleshy appendage ideal for the animal's sinuous swimming. All eels lack pelvic fins. Some eels have pectoral fins, but morays don't.

off-white, yellow, black and blue — with skin that can be speckled, striped, even polka-dotted. Morays are well equipped to blend with their surroundings; for some this camouflage even extends to the inside of the mouth. Their poor eyesight is compensated for by a keen sense of smell, which many use under cover of night to locate their prey. Others prefer to ambush prey from their burrows.

In certain tropical regions, the larger morays concentrate a naturally occurring toxin in their tissues, and the flesh becomes a dangerous delicacy. Our subject may have even tweaked English history in 1135. Following an eel-y meal, King Henry I succumbed to what is thought to have been food poisoning. Some eel-inclined scholars lay the blame not on the moray but on a "surfeit of lampreys," and both entrees are feasible given their availability at the time.

Into the reef's nooks and crannies tucks the moray, having found the perfect resting spot — and a place from which to ambush inattentive prey such as octopuses, fishes, or crabs and other crustaceans. You can count yourself lucky if you witness such a lightning-fast event, which can also occur as morays slither about on the hunt, either on the reef or in seagrass beds. Follow one sometime and see what happens!

DAVID FLEETHAM / ALAMY

This Undulated Moray (Gymnothorax undulatus) has captured a surgeonfish on the coral reef. Predation like this is not often witnessed.

Even if you see the moray snag its victim, what you won't be able to see is a second set of jaws that lie deep within the creature's throat, or *pharynx*. The eel projects these pharyngeal (fah-RINJ-ee-al) jaws forward, to latch

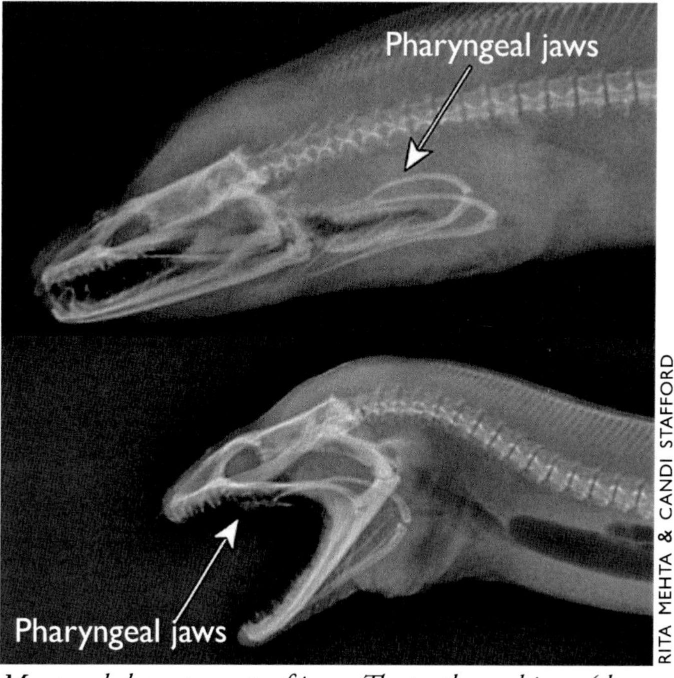

RITA MEHTA & CANDI STAFFORD

Moray eels have two sets of jaws. The toothy oral jaws (the heavier set shown here) make the initial grab. Then the pharyngeal jaws, equipped with their own set of teeth, shoot forward and help pull the prey into the eel's throat.

onto the prey and pull it farther down the throat, a frightening prospect, and the best reason to forgo poking one's fingers into the reef's cavities.

The mysterious ways of the moray helped it to arrive at supernatural status among ancient cultures. The Romans enjoyed dining on them, though they seem also to have been quite fond of some, taming them to respond to voice commands and decorating others with gold rings passed through the gills.

In his *Historia Animalium (History of Animals),* Aristotle suggested that eels were sexless and arose from the "entrails of the seas." (Think about that for a minute — it's the stuff of nightmares.) Three centuries later, the Roman naturalist Pliny the Elder posited that they reproduced by rubbing against rocks, loosening fragments that would then become mature eels. His book *Naturalis Historia* (*circa* AD77), is one of the largest single pieces of literature from the Roman empire. Its confident, oft skewed entries were a harbinger in a pre-internet world of how easily misinformation could be disseminated. (The world might have been enriched with more of Pliny's letters had he not found himself in Pompeii at exactly the wrong time.)

On the other side of the planet from our friend Pliny, Polynesian societies trace the origin of the coconut tree to the relationship between the female deity Hina and a spirit in the form of a moray eel. Hina raised the eel until it became so large that she was forced to release it into the wild. But the eel, smitten by now with its caretaker, bit Hina and followed her wherever she went. Fearful that the eel had become a threatening and cruel deity, villagers fed it a toxic potion that it greedily imbibed. Realizing his mistake, the eel called to Hina, "Since I am dying, let us part in peace. When you hear that they have cooked me, ask the head as your share. Take it and bury it, then look well after the plant that grows there. After three years,

TARA BONVILLAIN

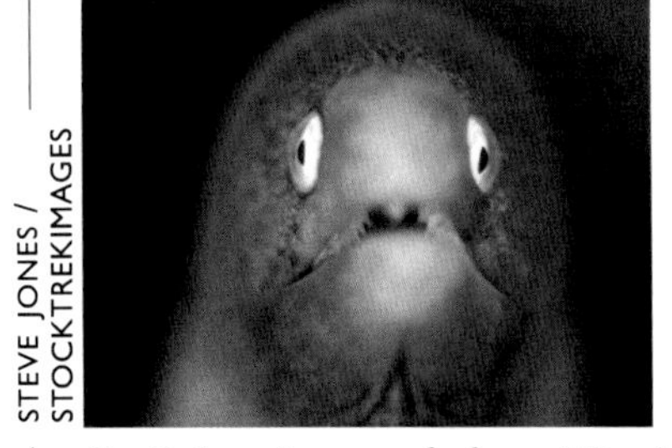

STEVE JONES / STOCKTREKIMAGES

In Polynesian mythology, Hina's eel gave her the coconut palm as a parting gift, to remind her of him.

you will see its bunched fruit. In the nuts you will see my eyes and mouth, and so we shall still be able to look at each other face to face. Pierce the largest of the three eyes. Drink the liquid and eat of the flesh of the fruit, as it is good. The name of the plant will be 'coconut.' All of its parts will be useful. The leaves will be a shade for you, and you will be able to plait them into mats and fans."

Hina's little fling with the moray is difficult but not impossible to top. In the real world of the coral reef, recent discoveries have demonstrated "social" links between some species of morays and groupers, as described below in "Hunting Buddies."

HUNTING BUDDIES

First observed in the Red Sea, the team-hunting behavior of the moray eel and the grouper, both of which normally hunt alone, involves a coordinated effort. The hunt might be initiated by a vigorous head shake on the part of the grouper in front of the moray's den, as if to say "I'm hungry, let's go try our luck." The moray slithers out of its home and they head off along the reef as if out for a casual stroll. The two exploit different hunting strategies: The moray frequently enters reef crevices while the grouper hunts in open water or ambushes its prey. As an efficient tag team, they reduce the escape options for prey. If the grouper detects prey hiding in a crevice, it signals the eel. If their victim spooks from the crevice, the grouper is there to pick it off. When the hunt proves successful, apparently the rule is first come, first served. There is no evidence to suggest the two enjoy a shared dinner date.

WATERFRAME / ALAMY

A Giant Moray (Gymnothorax javanicus) and a Leopard Grouper (Plectropomus pessuliferus, upper left) observed hunting together on a coral reef in Ras Mohammed National Park, Sinai, Egypt

9
Your Place or Mine?

In this Kingdom most of the plants are animals, the fish are friends, colors are unearthly in their shift and delicacy; here miracles become marvels, and marvels recurring wonders.

— William Beebe, *Half Mile Down,* 1934

Wedged into his bathysphere, William Beebe had descended to a record depth of 923 meters in August, 1934. On that dive off Bermuda, he observed marvelous new forms of life; Serpent Dragonfish, Sabre-toothed Viperfish, Scimitar-mouths and other magical names populate his notes of the day.

Two years later the peripatetic naturalist found himself in much shallower water off the tip of Baja California. He didn't need the bathysphere for this research dive, but he did need something that would allow him to spend some time at depth. Self-contained-underwater-breathing-apparatus, or SCUBA, was still a few years away.

To survey Baja's seabed, Beebe wore a diving helmet that was tethered to a surface air supply. It was a technique he'd first used in 1925 to view coral reefs in the Galapagos Islands, the first trained scientist to use the equipment. The helmet was heavy and awkward, but provided Beebe with the necessary time to observe astonishing creatures in action, edging science that much closer to understanding reef ecosystems.

Published in 1938, Beebe's account of the expedition, *Zaca Venture,* recorded wonder-inducing observations. Of one dive he wrote, "Before me, covering the considerable expanse of sand within view, was a garden of eels. . . . They resembled iron rods as much as anything, slightly bent above the middle of their length, sticking up from the sand, but, unrodlike, swaying very slightly. As I crawled slowly toward them, those nearest me, without

SOURCE: WILDLIFE CONSERVATION SOCIETY ARCHIVES

William Beebe taking notes underwater during a dive in Haiti in 1927. He breathed through a diving helmet with air supplied by a hose from the surface. (Reproduced by permission of Wildlife Conservation Society.)

effort... sank gently into the sand, until only the heads were visible. When even the latter disappeared, the sand closed over the small holes and the eels were as if they had never been."

Beebe was the first to link what seem to be incongruous words — *garden* and *eel*. But the label is accurate — and has stuck, for at a distance a colony of garden eels appears to be nothing more than a thin field of seagrasses, swaying back and forth before an underwater breeze. Peering down from the surface, it's easy to dismiss the "seagrass," and one's eyes move elsewhere in search of more interesting sights.

But submerging to the eels' level, preferably with scuba gear, will provide better perspective. Moving nearer to the "seagrass bed," the diver may think she's experiencing a sort of rapture of the deep, as the seagrasses begin to disappear tail-first until *pffft!* they're gone and the only thing left is a sandy flat, seemingly devoid of life. "Oh my, there's some more over there," she thinks and fins in that direction, only to experience the same spectacle. Odder still, there seems to be a sort of wave effect — the grasses in front disappear and those left in the diver's wake reappear! If our diver lets herself sink to the bottom and simply watches for a short time, she'll see the seagrass slowly sprout, a sort of time-lapse view of a new plant's emergence. But what plants are these? Plants don't have eyes!

If coral reefs are submerged cities, then garden eels live in the suburbs, choosing to inhabit sandy flats and slopes adjacent to the reef. Colonies with up to several thousand eels are possible, their permanent burrows spaced several centimeters apart like a submerged subdivision of single-

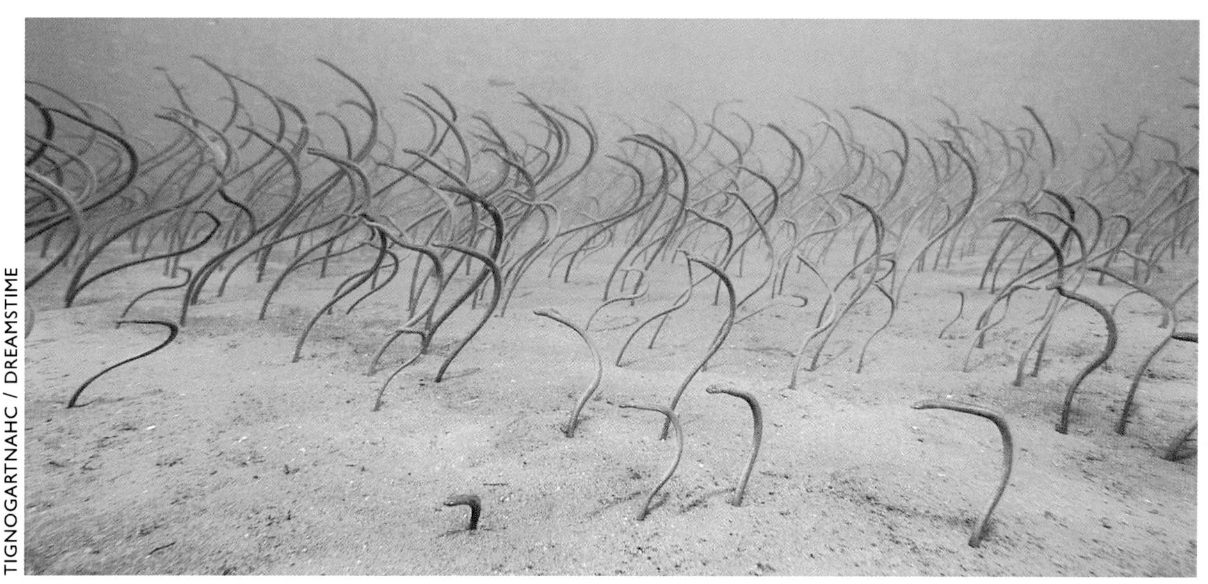

TIGNOGARTNAHC / DREAMSTIME

A garden of eels. Splendid or Orange-barred Garden Eels (Gorgasia preclara) live on sandy bottoms in the Indo-Pacific region.

resident homes. There is safety in these numbers, and as our diver found out, the eels react swiftly, cued by their neighbors to periscope down, tail first, in response to a threat.

Garden eels are secure from all but the most persistent predators. The Pacific Snake Eel is one of those predators. Much larger and more powerful than a garden eel, the Snake Eel burrows into the sand near a colony, digs a tunnel under an eel and grabs it by its tail. Stingrays are less cunning and use rapid undulations of their powerful fins to dig the eel out. Large triggerfishes huff and puff, blowing bursts of water at the eel's lair until the creature is exposed and can be grabbed and eaten.

Garden eels are found worldwide, and all 35 or so species are pencil width and about 50 cm long. Their colonies may be seen at various depths but never so deep that the darkness makes it difficult to spot prey. The eels bend into the current, feeding on tiny crustaceans known as *copepods* (KOH-peh-pods) that drift past. But no matter how tempting it might be to swim after a tasty tidbit, the eels remain anchored to their burrows, true homebodies, rarely leaving except to establish a new home at mating time.

An unusual lifestyle like this requires some specialized anatomical adaptations. The caudal (or tail) fin of a garden eel is modified into a hard fleshy point, which the eel can drive into the sand. As the resulting burrow runs deeper, the eel's dorsal fin comes into play, funneling the sand auger-like up and out of the hole. Cave-ins are averted by cementing the walls of the burrow with slime produced by a gland near the tip of the tail as the eel moves deeper into the sand.

As mating season approaches, garden eel thoughts turn to construction requirements. Cohabitation in a structure built for one isn't possible, so males and females build new burrows close enough to each other to consummate the act, an entwining ballet performed while each partner remains anchored to its home port. Fertilized eggs are released into the current, floating close to the surface, bound for the open ocean. Most eggs and juvenile eels are snapped up by hungry mouths, but a few survive and drift along until large enough to swim down to the seabed, there to tunnel in and populate another suburban enclave.

TOBIAS BERNHARD RAFF / BIOSPHOTO

A pair of Spotted Garden Eels (Heteroconger hassi) courting

10

Breathtaking!

The commonest habit of sea cucumbers when annoyed is to discard practically all their internal organs, then to wait patiently until time has regenerated a new set, so that they may again eat, drink, and bring about whatever a holothurian can in the way of merriment.

— William Beebe, *Zaca Venture*, 1938

Though bathers may think of them as a mushy hindrance to a good dip on a sandy bottom, the sea cucumber is well worth a careful look, if for no reason other than to admire its variety of preposterous habits. Living vacuum cleaners that range from a few centimeters up to 3 meters long, the leathery, lumpy sea cucumbers slowly make their way along the seabed, slurping up bits of organic food particles with a tentacle-lined mouth, and sand along with them. The creature is so busy eating it seems never to take time out for a breath. But that's because it doesn't have to! Food gathering and breathing take place at opposite ends of this creature. If so inclined, one can observe the sea cucumber's anus as it opens and closes, transporting oxygenated water into the body. But wait, there's so much more going on back there.

Members of a class of organisms known as Holothuroidea (HOLL-uh-thur-OID-ee-ya), sea cucumbers are sometimes loosely referred to as "sea slugs." But unlike the others called sea slugs (on page 20), they are not mollusks, not at all related to snails or terrestrial slugs. They are in fact ancient *echinoderms* (ee-KYE-nuh-derms), relatives of starfishes and sea urchins that have been lumbering about for some 400 million years. Highly adaptable, they

CAROLINE S. ROGERS

Taking its name from the three rows of tube feet (hydraulically operated suction cups) on its underside, the Three-rowed Sea Cucumber (Isostichopus badionotus) "processes" as much as two tons of sand per year while feeding virtually nonstop.

occupy a range of habitats including seagrass beds, coral reefs, and sandy and muddy bottoms. Shallow water is suitable, and so too are deep canyons like those plumbed by daring oceanographers starting with William Beebe in the 1930s. Sea cucumbers have their share of wonderfully descriptive names, such as the Pineapple, the Tiger Tail and the ever popular Donkey Dung.

CAROLINE S. ROGERS

Covered with hundreds of soft tube feet on both dorsal and ventral surfaces, the Furry Sea Cucumber (Astichopus multifidus) looks and feels exactly as named.

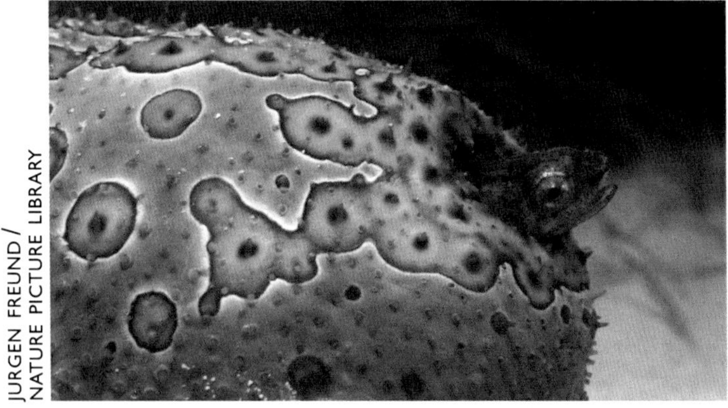

JURGEN FREUND / NATURE PICTURE LIBRARY

Without much say in the matter, a spotted sea cucumber hosts a pearlfish (Carapus sp.), which may have more in mind than just couch surfing.

When it inhales, the sea cucumber may be in for a bit of a surprise. Another reef resident, the slender, dagger-shaped pearlfish, sees a holothurian's anus as a promising bit of real estate. Tail-first and without asking permission, it snuggles into the nether niche, leaving only at night to forage. In the natural world, such a "landlord-tenant" relationship is termed *phoresis*, an association in which one organism attaches itself to a "host" strictly for the purpose of getting from place to place. But a model phoretic the pearlfish is not, and in a rare example of a vertebrate parasitizing its invertebrate host, this little fish may occasionally snack on its landlord's reproductive organs.

Still other breathtaking activities occur at this bizarre terminus. As noted by Beebe, some sea cucumbers employ a unique defense strategy. Attached to their respiratory centers are the *Cuvierian tubules* (koo-vee-AIR-ee-yan TOOB-yools), a gizmo that even Spiderman would

CBIMAGES / ALAMY

An annoyed Brown Sandfish sea cucumber (Bohadschia vitiensis) discharges its Cuvierian tubules in self-defense.

envy. Belched out when the creature feels threatened, the tubules become sticky on contact with water and quickly entangle the hungry enemy, be it a crab, lobster or fish. Meanwhile, the sea cucumber "scurries" away on its hundreds of hydraulically powered tube feet. As if that wasn't enough, the tubules sometimes contain a poison, *holothurin* (holl-uh-THUR-in), used by Pacific Islanders to anesthetize food fish in enclosed tide pools.

As a food source, sea cucumbers are less important to island nations as a dietary staple than as a money maker. A thousand years ago, the Asian appetite for them introduced many Pacific Islands to world trade networks. Sold in their dried form as *trepang* or *bêche de mer*, they are used to flavor soups, stews and stir-fries, though the Western palate might find them a tad off-putting. One such diner once rated the taste as "slightly lower than phlegm, the texture of which it closely resembles."

Sea cucumber, or beche de mer, *can be found fresh, salted or otherwise preserved in seafood markets like this one in New York City's Chinatown.*

Medicinally, sea cucumbers are used as treatments for such ailments as high blood pressure, joint pain and naturally, a depressed libido. And we can add them to the list of reef creatures with intriguing pharmacological value. Turns out they make a protein, *lectin,* that may be useful in genetically engineering mosquitoes to prevent them from hosting malarial parasites and passing the disease to humans.

Sea cucumbers are partially responsible for giving birth to a new tongue. *Bislama* is a mutated construction of the Portuguese *bicho do mar*, meaning "small sea creature." The Bislama language is an English-based pidgin, a simplified language born of the need of late 19th-century traders to communicate with their Pacific contacts engaged in the sea cucumber trade. In his epic book about his South Seas adventures, *The Cruise of the Snark*, Jack London called it "a language so simple that a child could learn it," with a very limited vocabulary. So the next time a case of *mal de mer* announces itself following a meal of *bêche de mer,* one of London's Bislaman aphorisms might come in handy: "Belly belong me walk about too much."

11
The Fish That Fishes

> *... the fishes included in this collection were drawn and painted by me... from nature. This was done to the best of my ability, not believing that human arts can express the beauty of the colors of these fishes when caught alive in the best season or when we can see them swimming in clear water.*
>
> — Samuel Fallours, letter to Louis Renard, 1718

At the beginning of the 18th century, fishermen from the island of Ambon in what is now Indonesia brought creatures from the coral reef to the Dutch artist Samuel Fallours. Part-time artist, full-time soldier, Fallours was employed by the Dutch East India Company, the powerful organization that held sway over the islands, exploiting them for the lucrative spice trade. At the time, Ambon was just becoming known for its lush nature, including the reef life, and Fallours had many models to choose from.

His style was radical for the time, bordering on surrealism in its interpretation of the natural world. His whimsical, brilliantly colored fishes were the stuff of dreams, decorated with suns, moons, stars and potted plants. His spiny lobster climbs trees and lives in the mountains. Tiny human faces peer up from the carapaces of his crabs. We are denied the source of his exquisitely rendered mermaid, but we're teased by his claim that he kept the "siren alive for four days in my house at Ambon in a tub of water."

The mermaid and over 400 other illustrations found their way onto the pages of one of the rarest natural history books of all time. Arriving in 1719, Louis Renard's *Fishes, Crayfishes and Crabs* was the first color "guidebook" of sea creatures. One rendering was of a very odd specimen, referred to as the "Sambia," "Walking Fish" or "Common Fish" of Ambon.

But let's back up a few years to late December of 1696, when off the coast of West Australia, another Dutchman, Captain Willem de Vlamingh and his vessel *Geelvinck* searched for the victims of a previous year's shipwreck. Surely they were distracted from their task on more than one occasion, encountering "rats as big as common house cats" (no doubt wallabies) and "tracks of tigers" (perhaps dingos or *thylacines* (THIGH-luh-seenz), marsupial wolves now thought to be extinct). And then there was a remarkable fish "about two feet long, with a round head and a sort of arms and legs and even something like hands." The fish was probably Fallours's *Sambia.* We know it as the frogfish, arguably one of the oddest creatures on the coral reef, if not the planet.

ARTIST IN RESIDENCE

Artist Samuel Fallours painted for profit during his tenure on Ambon Island in 1718, his renditions executed with what seems to be little attention to scientific precision. His work appears at first glance to be of minor scientific value, a quaint peek into the natural history of the period. Yet 300 years later we know that most of his illustrations are renditions of actual species, and they are now viewed as a part of the scientific record for the day. He had little interest in being exact, apparently settling on a colorful, innovative style that he believed would lure wealthy, eccentric collectors to his paintings. And on occasion he even upped the ante, captioning them with lines that fracture the imagination. Of the Sambia, he leaves us with this: "I caught it on the sand and kept it alive in my house for three days; it followed me everywhere with great familiarity, much like a little dog."

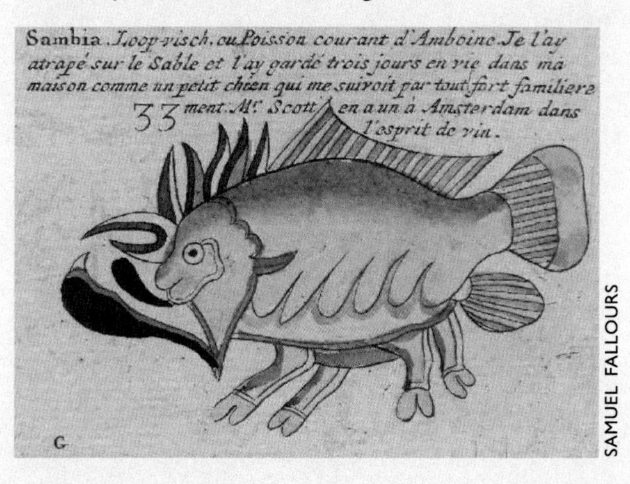

SAMUEL FALLOURS

Frogfishes come in a wide range of stunning colors and are found in nearly all tropical to temperate oceans and seas. Ranging in size from thumbnail to a slightly squished soccer ball, they have been described as globose, chubby and ugly. Their appearance, combined with all their weird unfishlike traits, makes them one of the most appealing species of fishes an ichthyologist can study. Just imagine specializing in a family that includes the Sargassum Frogfish, a creature quite content to spend its life drifting in a mat of seaweed. Then there are the Warty, Side-jet and, not to be outdone, Tail-jet frogfishes.

CAROLINE S. ROGERS

Previously believed to reside only in floating mats of Sargassum algae, Sargassum Frogfish (Histrio histrio), 20 of them, were discovered in 2010 among the shallow prop roots in the mangrove-fringed bays of St. John in the U. S. Virgin Islands. The hurricanes that struck the islands in September, 2017, severely damaged this habitat, and the return of this wondrous fish is uncertain.

And like a long-lost flamboyant uncle who shows up out of nowhere, we have the recently identified (2009) Psychedelic Frogfish, so named for its mind-bending kaleidoscopic patterning. The Psychedelic's eyes appear to be directed forward, allowing their radii of vision to overlap. Presumably, this would provide the fish with some level of depth perception based on a certain amount of binocular vision, a characteristic few fishes can brag about.

When one first spots a frogfish, there may be an overwhelming urge to giggle uncontrollably. "Surely there's a mistake," the brain comments. "Fishes aren't supposed to look like this!"

Amazement soon follows, as the viewer notes how well such a strange little creature blends in with its surroundings. Using a combination of warts, filaments, stripes, spots and skin flaps, frogfishes can be structurally and chromatically cryptic. They look like rocks, seaweed, coral, even creatures such as sponges or tunicates, anything but a fish. It's a strategy of camouflage known as *aggressive resemblance,* because it signals an interesting background to potential prey animals, which mistake the frogfish for a possible shelter or foraging locale.

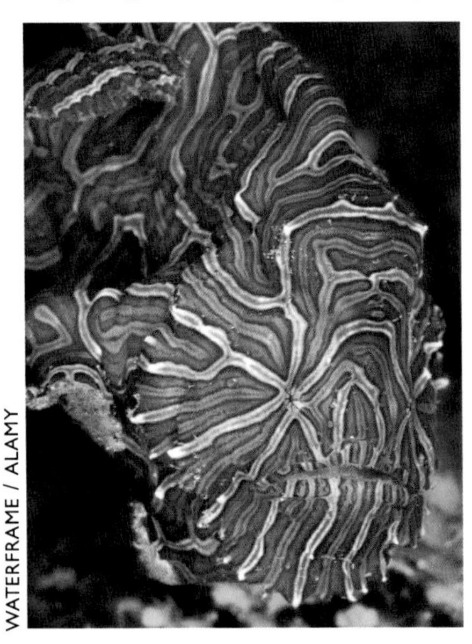

WATERFRAME / ALAMY

With its peculiar whorled pattern of stripes, it's easy to see how the Psychedelic Frogfish (Histiophryne psychedelica) got its name.

A well-equipped frogfish is essentially an angler geared up with rod, lure and lots of patience. One of nature's most advanced "lie in wait" predators, the frogfish uses a strategy that has fascinated scientists for centuries. In *Historia Animalium* (fourth century BC), Aristotle noted that "the fishing-frog has a set of filaments that project in front of its eyes ... and are used as baits ... the animal ... raises the filaments, and, when the little fish strike against them, it draws them in underneath into its mouth ... ." The filament or "fishing pole" he mentions is a modified dorsal spine, the *illicium* (il-LISS-ee-yum), and is the source of the family name, Antennariidae (an-teh-NARE-ee-uh-dee). The illicium is tipped with a lure known as the *esca*, which is unique to each species. In the film *SpongeBob SquarePants,* the esca was an ice cream stand connected to a giant frogfish that nearly ended Sponge Bob's escapades. In the real world, some escae resemble miniature pom poms, others look like shrimps; but all are what biologists refer to as an "aggressive mimetic device." As prey approaches, like any smart fisherman the frogfish waggles the lure (or in SpongeBob's case, offers a free ice cream cone) to draw its victim to within striking range.

So it is that most frogfishes spend a good deal of time hanging out, waiting for the next snack to wander by.

ORLANDIN / DREAMSTIME

A Hairy Frogfish (Antennarius striatus), with its wormlike esca and waiting maw

But don't let this couch potato–like behavior fool you. Frogfishes are among the most talented and voracious of predators and aren't at all picky about what they eat. A Sargassum Frogfish once was found to have 14 of its brethren in its stomach! Little seems to escape their gastronomic attention, and with a mouth capable of expanding to 12 times its normal size, the frogfish can inhale something twice as big as itself. Here's the kicker; the feeding action lasts — hold on to your fork — just one-fiftieth of the time it takes to blink your eyes.

Like all fisherfolk, frogfishes are quite attached to their favorite spots and don't need to move about all that much. But if the urge strikes, off they go. Kitted out like a submerged ATV, the frogfish uses its pectoral and pelvic fins to lumber across the seabed over rocks, coral and whatever else gets in its way. Should it hit an impasse or need to vault a chasm — not a problem. Like squids and octopuses, minus the poise, frogfishes have the ability to suck in water, then eject it — from small openings behind the gills. Jet propelling itself in fits and starts from point A to point B, the frogfish is the reef's awkward child on a stage replete with creatures of grace and agility.

WHITCOMBERD / DREAMSTIME

SUWAT SIRIVUTCHARUNG-CHIT / DREAMSTIME

JEREMY BROWN / DREAMSTIME

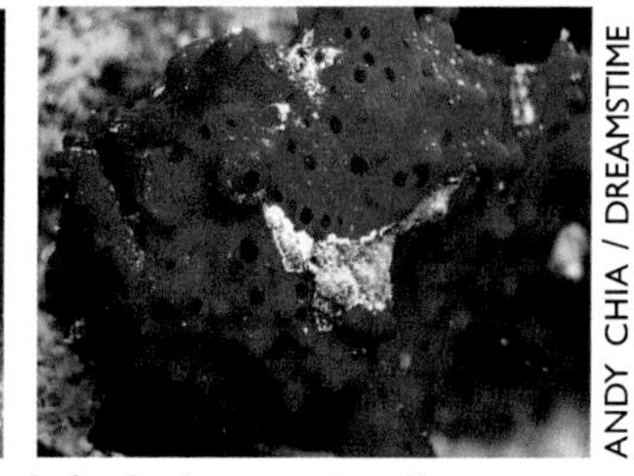

ANDY CHIA / DREAMSTIME

A few more frogfishes, from top left clockwise: A yellow Warty or Clown Frogfish (Antennarius maculatus), a white Clown Frogfish (also Antennarius maculatus), a Red Frogfish (Antennarius nummifer; its mouth is over here on the left side) and a yellow Longlure Frogfish (Antennarius multiocellatus)

12

Armed and Curious

Octopuses practically beg to be mythologized. Long before *Jaws*, there was the *kraken*, a catch-all term for giant octopuses and squids bent on wreaking havoc. Symbol of evil in print, illustration and film, the octopuses — for centuries known as devilfish — occupied the dark alleys of every mariner's imagination. In one particularly vivid drawing he called *Poulpe Colossal,* French zoologist Pierre Denys de Montfort brought every sailor's nightmare to life in his 1810 woodcut of a bug-eyed octopus overpowering a sailboat. The work was based on stories told by whalers who claimed to have seen evidence of such giants at sea.

The octopus in John Steinbeck's classic novel *Cannery Row* is a "creeping murderer" with "evil goat eyes." Victor Hugo chillingly called it "one of those embryos of terrible things that the dreamer glimpses confusedly through the window of night." In exile on the island of Guernsey, Hugo let his imagination spin freely, sending Gilliatt the fisherman to battle an enormous octopus in *Les Travailleurs de la Mer* (*The Toilers of the Sea*). "Compared to the devilfish," he wrote, "the hydras of old bring a smile to the lips."*

In the mid–20th century octopuses crept onto the pages of pulp fiction, where they proceeded to engulf superheroes, divers and damsels in distress. Hollywood promoted them as both gargantuan and diminutive shiver-inducing creatures of the silver screen. A Golden Gate–sized specimen snacks on San Franciscans in the 1955 thriller *It Came From Beneath*

BARRY B. BROWN / WILD HORIZONS

A charismatic Reef Octopus (Octopus briareus) floats in the dark near Curacao, Netherlands Antilles.

*The mythological Greek Hydra that Victor Hugo referred to was a fearsome nine-headed marine monster.

Poulpe Colossal, *by Pierre Denys de Montfort, 1810*

the Sea. Even Disney found the image of a wicked octopus irresistible, creating the nefarious Ursula opposite the winsome Ariel in *The Little Mermaid.* Leading men from King Kong to Ian Fleming's agent 007 have had to contend with deadly octopods. In *Octopussy,* James Bond kept tabs on his femme fatale's tiny pet blue-ringed octopus, whose toxic bite can lead to death within minutes — and, of course, did for one of Bond's pursuers.

But in a charming coastal Italian town, at least one octopod gets a pass. Legend has it that a giant octopus once crept out of the Ligurian Sea and rang the church bell of Tellaro, alerting villagers to marauding pirates headed their way. Of course, folk hero status only goes so far, and *polpo* is eagerly consumed throughout the town.

The science of octopuses is equally as exciting as the mythology. It may be hard to believe, but octopuses are mollusks, distantly related to shellfish such as clams, oysters and snails. Over a hundred million years ago, an evolutionary road split; the ancestors of clams went one way, snails and slugs another. The precursor of today's octopuses took still another genetic highway; trading the security of a shell, they became specialized for speed, mobility, keen vision and star wattage as an aquarium draw. Octopuses are kin to squids, cuttlefishes and the rarely seen nautilus. All are cephalopods (SEFF-uh-lo-pods), members of the class Cephalopoda (seff-a-LA-puh-da), a word that tells it like it is — "head-footed."

With a classy name like Cephalopoda, how can you not find them marvelous? They crawl about the seabed on eight arms lined with individually controlled suction cups. Or jet around by drawing water into their *mantle* —

PICS516 / DREAMSTIME

A Greater Blue-Ringed Octopus (Hapalochlaena lunulata) on a reeftop in Lembeh Strait, Indonesia. Blue-ringed Octopus are beautiful, deadly and small (some species are only 12 cm from arm tip to arm tip across the body).

the muscular, bulbous head flap containing the creature's organs.Forcefully expelling the water through a funnel at the side of the head propels the octopus from place to place, speeding it off to safety. Their suckers are lined with chemoreceptors that can taste food. Octopuses solve problems, use tools, learn skills and recognize individual humans. Their hearts are not one, not two, but three. If frightened, they jet away behind the cover of an ejected cloud of black ink.

Found in all oceans, there are several hundred species of octopuses. The Pygmy is thimble-sized, typically weighing in at merely 30 grams. The Giant Pacific Octopus can grow to a scale-busting 150 kg or so. Other octopus species include the Bumblebee, Flapjack, Gloomy, Old Woman and Wonderpus octopuses as well as the darling of Caribbean snorkelers, the commonly seen Common Octopus and the anything but common Caribbean Two-spot Octopus.

All octopuses have a level of intelligence otherwise unheard of in the realm of invertebrates. One quickly learned to solve a maze by watching a trained octopus navigate the puzzle first.

Veined Octopus collect shell halves — clam or coconut — for use as shelter. When threatened, they tuck into one half and may even pull the other on top and wait for the danger to pass. They have been seen transporting the shells from one place to another, using them as "mobile homes."

One thing octopuses don't do very well is to live long lives. Programmed to grow fast, breed once and die young, the oldest of them are but three or four years old.

"THE OCTOPUS" by Ogden Nash

Tell me, O Octopus, I begs,
Is those things arms, or is they legs?
I marvel at thee, Octopus;
If I were thou, I'd call me Us.

ALEX MUSTARD / NATURE PICTURE LIBRARY / ALAMY

A Veined Octopus (Amphioctopus marginatus) has crawled into half a coconut shell and appears to be in the process of pulling another half shell over the top, perhaps in response to the photographer.

As we've learned more about these complex beings, they've become less threatening and more fascinating. In his 1940 essay, "In Defense of Octopuses," the naturalist Gilbert Klingel wrote, "I feel about octopuses — as Mark Twain did about the devil — that someone should undertake their rehabilitation." Soon after that, Jacques-Yves Cousteau brought octopuses into our living rooms via television. Today, rather than feared, the octopus is on many a reef visitor's must-see list. Ah, easier said than done.

Looking for an octopus on a coral reef is like seeking the proverbial needle. During the day an octopus often holes up in its den: a small cave or crevice, even a beer bottle will do (preferably stubby with brown glass, according to one observer). But at night, many octopuses begin their predatory rounds. Armed and dangerous like its cousin the squid, an octopus is an efficient, voracious

CAROLINE S. ROGERS

The rare Caribbean Two-spot Octopus (Octopus hummelincki) here displays its bright-blue "eye" spots as it shelters among the spines of a Long-spined Sea Urchin (Diadema antillarum). (Note the two fireworms, distant relatives of the earthworm, perhaps also sheltering among the spines. If you ever see one, don't touch it; they sting.)

WHY THE OCTOPUS HATES THE RAT

As the story goes, after its canoe sank, the rat was offered a ride to shore by a passing octopus. The rat happily jumped atop the octopus's head, was shuttled to a nearby island and after hopping to safety, shouted, "Thanks, Octopus, I left you a little present on your head!"

With an arm, the octopus reached up and discovered that the rat had left a pile of excrement up there. Since then, whenever an octopus spots a rat, it goes into a rage. Knowing this, Pacific Islanders construct a ratlike stone-and-Leopard-Cowrie-shell lure, then dangle it over the octopus's lair. Spotting its old nemesis the rat, the octopus rushes forth and seizes the lure, only to be drawn to the surface by the fisherman.

ROGER BOWEN

A Maka Feke stone-and-shell octopus lure from the Tongan Islands

hunter. It stalks crabs, probes for clams, explores crevices for shrimp. Because each of the suckers on the arms functions independently of the others, octopuses can manipulate objects with ease, even to the point of rolling a tasty tidbit sucker-to-sucker on down to the mouth. There the creature dispatches its prey with a lethal bite, or if it's a clam not-yet-on-the-half-shell, pries it open. Should the bivalve resist, the octopus drills a hole in the shell using an oral structure called a *radula* (RAD-yoo-la), injects venom and waits for the appetizer to succumb. Often, the hunter retreats to the safety of its den to eat. Meal over, the octopus doesn't want to crowd up its living room, and like sloppy college students tossing pizza boxes outside a dorm room, an octopus discards shells outside its den. Want to find an octopus? Look for its garbage heap.

If you're lucky enough to spot an octopus caught out in the open, don't look away. One moment it's there, the next it's not. A master of camouflage, an octopus has many tricks up all those sleeves to avoid detection. It has thousands of skin *papillae* (pa-PILL-ee), protrusions that can be raised or lowered to create bumps, lumps and even horns, all very convenient as a way to match the textural world of the coral reef. Octopus skin also harbors an array of minute color organs known as *chromatophores* (kro-MAT-uh-forz). Like tiny cartridges of printer ink, the chromatophores can change the color in any part of an octopus's body in less than a fifth of a

CAROLINE S. ROGERS

This octopus has made its home in a concrete block. Note the discarded pen shells, delicate triangular shells of bivalves that are apparently a favorite food of this particular resident.

HOWARD CHEW / ALAMY

The Wonderpus Octopus (Wunderpus photogenicus) has long arms and a rusty brown color with white spots on its mantle and white bands on its arms. Like other octopuses it can change its appearance to camouflage itself or to mimic an aggressive reef dweller such as a lionfish.

second, an "app" that would be the envy of any fashion designer. Like to see this outfit in lavender stripes or turquoise polka dots? How about alternating psychedelic waves of maroon and cream? Need to look bigger than you are? Just add dark "makeup" around the eye, and *voilá* — you've got a really big eye! Just a few of octopuses' disguises mimic such things as stationary rock, waving seaweed, a sedentary sponge or a brain coral. Even the flattened flounder isn't off limits.

With all its illusionary powers to avoid detection, one threat looms large in octopus gardens worldwide. As levels of carbon dioxide in the atmosphere climb, ocean pH is falling. Octopuses' copper-based blood makes these creatures less tolerant than most of their reef neighbors to increasingly acidic seas. Ultimately, they may be the reef's "canary in the coal mine," and future reef visits may include the sight of empty dens, waiting for new occupants that won't be coming.

THE EYES HAVE IT?

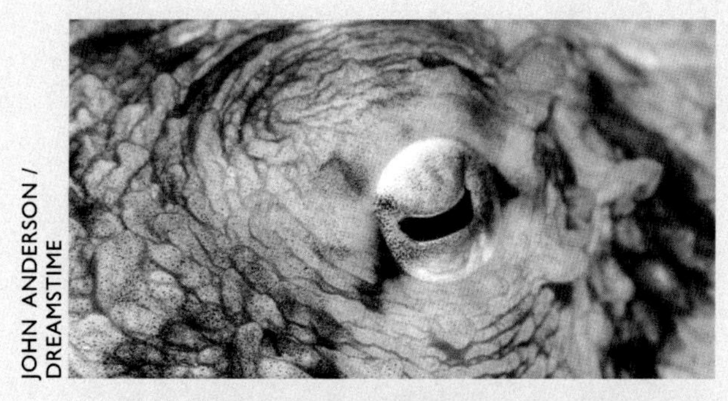
JOHN ANDERSON / DREAMSTIME

CAROLINE S. ROGERS

What makes octopus camouflage so remarkable is one little factoid — the octopus eye has only one type of visual pigment, the kind that senses light and dark (black, white and shades of gray only, no color), so we would expect the animals to be color blind. But their U-shaped or dumbbell-shaped pupils may play a role in separating the wavelengths in light to make it possible for octopuses to detect colors in an entirely different way than humans and other animals with color vision do, an ability that might play an important role in their camouflage.

Alexander Stubbs, a biologist from University of California, Berkeley, and his father, astrophysicist Christopher Stubbs from Harvard University, have investigated how the oddly shaped pupils and movable retinas of octopuses (top) and other cephalopods (such as the Caribbean Reef Squid, bottom) might allow those species to detect color by separating the wavelengths of light, providing color vision without the three or more color pigments present in the eyes of other animals that see colors.

13
Reef Butterflies

Whoever thought of the name "butterflyfish" probably didn't have to work very hard at it. Trading wings for fins, a butterflyfish does seem to resemble the insect from which it takes its name as it flits from coral to coral in search of tasty invertebrates and bits of algae. Its teeth are well suited for such fare and give a clue as to the origins of the family name, pairing *chaeto* with *odont* to make Chaetodontidae (kee-toh-DON-tih-dee), or hair-toothed.

These are laterally compressed disc-shaped creatures, ichthyology-speak that translates as a fish designed like a pancake. The shape allows a butterflyfish to slip easily into the reef's narrow passageways, even horizontal ones by swimming on its side.

On a healthy reef, especially in the Pacific or Indian Ocean, chaetodontids (kee-toh-DON-tids)are some of the most common and conspicuous fishes. Like their terrestrial namesakes, they come in a variety of gold, black and orange hues accented by dazzling patterns that seem straight out of a surrealist's dreamscape. Their names, almost equally dazzling, include the Sunburst, Vagabond, Longnose and Exquisite butterflyfishes. One, the Wrought-iron Butterflyfish, is said to remind some people of traditional samurai kimono wear because of its black-and-white patterning.

SEADAM / DREAMSTIME

A school of Four-eyed Butterflyfish (Chaetodon capistratus)

DIVEDOG / SHUTTERSTOCK

The Wrought-iron Butterflyfish (Chaetodon daedalma)

In the coral reef world, committed relationships are rare. Fishes have enough trouble managing their own lives without worrying about how well anyone else is doing. Most chaetodontids, however, have adopted an alternative lifestyle and seem to do quite well as monogamous pairs, sharing and defending territories. Without a doubt, two sets of eyes must also be advantageous when on the lookout for predators.

POELZER WOLFGANG / ALAMY

STEPHANKERKHOFS / DREAMSTIME

The Yellow Longnose Butterflyfish (Forcipiger flavissimus, top) and Crown Butterflyfish (Chaetodon paucifasciatus) are two of the many spectacular butterflyfishes.

Because they are a popular menu item for many reef predators, some butterflyfishes have become masters of deception. The Four-eyed Butterflyfish, a favorite of Caribbean snorkelers, sports a large dark spot on either side of the tail. This extra pair of false eyes is more conspicuous and therefore more likely to be the point of attack for predators. A black vertical stripe disguises the true eye, which makes it difficult for predators to make heads or tails of their potential meal. When a hungry moray eel goes for the tail end, our little friend darts off, leaving the puzzled eel with nothing more than a mouthful of water.

Our four-eyed fishy friend, *Chaetodon capistratus,* first described in print in 1758, owes its scientific name to Carl

PETER LEAHY / DREAMSTIME

To get the full effect of the deception, we can use the old graphic designer's trick of squinting at this photo of the Four-eyed Butterflyfish to mute the fine detail. Then go back and squint at the school on page 56.

Linnaeus, one of the most influential scientists of his time. Linnaeus developed our modern system of classifying and naming plants and animals, a discipline known as *taxonomy.* "God created, Linnaeus set in order" is the catch phrase for this Swedish luminary's *binomial nomenclature* system. And yet, since Linnaeus never once experienced the heat of the tropics, how did the original scientific descriptions of butterflyfishes and other reef species end up being attributed to him?

The trail leads us back to 1729 and to Linnæus's best friend at Sweden's Upsala University, the equally gifted naturalist Peter Artedi. In the early 18th century, an amalgamation of life forms was showing up in Europe from all corners of an ever expanding world. Confusion reigned. Was a shark a fish? What to make of all those six-legged bugs, many of which looked alike but not exactly? There were jewel-like seashells, as well as fish skins that were like nothing ever seen before.

Many of these specimens, butterflyfishes included, resided in museums as well as in what were known as *Kunstkammern*, "cabinets of curiosities" proudly maintained by well-heeled collectors. But the specimens were often arranged with little attention to anatomical details, and they remained nameless or were merely assigned colloquially inaccurate labels. There was a hunger among naturalists for a logical scheme that would permit them to properly name and categorize this explosion of life, to make some sense of it all.

Rather than pieces of furniture, Kunstkammern, *or cabinets of curiosities, were often entire rooms. Their contents were curated to greater or lesser degrees, depending on the interests and abilities of their owners. "Musei Wormiani Historia," the frontispiece from Olaus Wormius's* Museum Wormianum, *published in 1655 after his death, depicts the cabinet of curiosities belonging to Danish physician Ole Worm (or Olaus Wormius as he called himself in Latin).*

Enter Linnaeus and Artedi. Rapidly, the two young men bonded over their common interest in designing a workable system to name all plants and animals. One can imagine them haggling over who would take on the amphibians, who the birds, insects and so on. Already an expert botanist, Linnaeus corralled plants; Artedi took the amphibians, the reptiles and, happily so for him since they were his passion from a very young age — the fishes. The two vowed to finish the work should death claim either one prematurely, little knowing the pact would be tested all too soon.

The years slid by but the work of a naturalist wasn't that financially rewarding and by 1735 Artedi showed up in Amsterdam, penniless but having made great progress with *Ichthyologia,* his compendium of fish classification that would one day earn him recognition as the father of ichthyology (ick-thee-ALL-uh-jee). Needing to locate more specimens, especially from the tropics, he found himself, courtesy of Linnaeus, in the midst of an astonishing collection of life forms, the cabinet belonging to a wealthy apothecary, one Albertus Seba. Seba acquired his marine specimens through a far-flung network of international contacts as well as from sailors just in from the tropics, eager for his much needed potions in exchange for their fishy souvenirs from afar. Some of the fishes he utilized in pharmaceutical applications, others found their way into his display collection, to be professionally illustrated and added to his growing *Thesaurus,* the finest and most comprehensive survey of life's diversity of its time. *Volume III* was dedicated to marine life, and to complete the work on fish, he turned to Artedi.

Picture a wide-eyed Artedi as he took in Seba's cabinet for the first time, scrutinizing sea urchins, squids, crabs and fishes the likes of which he must never have imagined.

SOURCE: KONINKLIJKE BIBLIOTHEEK / NATIONAL LIBRARY OF THE NETHERLANDS

Labeled "Tropical coral fish from the Caribban and Indo-Pacific, Plate 25," this page from Albertus Seba's Thesaurus (Volume III), *published after his death in 1736, depicts several species of butterflyfishes (indicated here by red dots ●). About a century later, French entomologist and editor F. E. Guérin found Seba's original copper engravings and republished the fish plates with new text and updated species names.*

There were porcupinefishes, triggerfishes, cowfishes, flying gurnards — and yes, butterflyfishes — each in need of classification and a Latin name that would establish it in the scheme of life. He worked diligently for 10 weeks, banking fish upon fish for inclusion in Seba's *Thesaurus* as well as for his own publication.

Then, on a September evening in 1735, Artedi decides to take a well-earned break from his labors. He accepts his sponsor's dinner invitation. Seba's food and excellent wine are welcome fare for one deprived of such luxuries. Knowing he is on the cusp of publishing his work, perhaps Artedi decides to celebrate and overdoes it a bit. He departs after midnight and, it is later presumed, in his disorientation he slips at a most inopportune moment. The next morning his body is discovered floating in a canal.

Honoring their agreement, Linnaeus gathers his friend's papers and notes and completes Artedi's work. By 1758 the Chaetodontidae and many other reef fishes are formally revealed to the world in the 10th edition of Linnaeus's opus, *Systema Naturae* (*System of Nature*). Rightfully or not, because of that volume Linnaeus is credited with describing many of the fishes that his friend had initially examined and classified. Artedi, for all his dedicated labor that ended on a slippery bank with a tragic death at the age of 30, is known but to a small audience as the person hidden in the shadow of his fellow countryman.

We should at least cast a sliver of light on the young man who perished all too soon, for absent both his and Linnaeus's work, the creatures and plants with whom we share the earth would be a mishmash of names no one could sort out:

> *Here lies poor Artedi, in foreign land pyx'd**
> *Not a man nor a fish, but something betwixt,*
> *Not a man, for his life amongst fishes he past,*
> *Not a fish, for he perished by water at last.*
>
> — Anders Celsius, 1735

*"Pyx" comes from the Greek word "pyxis," meaning box, or receptacle. Here "pyx'd" probably means "boxed" or "casketed." Artedi, who was Swedish, was buried "in foreign land," in an unmarked pauper's grave in Amsterdam, Netherlands.

14
Smile!

The great pack of barracudas seemed to have gone mad. They were whirling and snapping in the water like hysterical dogs.... The water was boiling with the dreadful fish and Bond was slammed in the face and buffeted again and again within a few yards. At any moment he knew his rubber skin would be torn with the flesh below it and then the pack would be on him.

— Ian Fleming, *Live and Let Die**

Great Barracuda lead their fear-free lives in the world's temperate and tropical waters. They can be found around reefs, above shallow sand flats and in mangroves, where juveniles tend to hang out before venturing to deeper water. While snorkeling among the mangroves, one researcher noted that even at a few centimeters long baby barracuda seem to have an "attitude," and are not likely to be intimidated. Barracuda are also fond of skulking about submerged wrecks, where they apparently find great sport in shadowing scuba divers. In more open

CAROLINSE S. ROGERS

It's difficult to mistake the Great Barracuda (Sphyraena barracuda) for any other reef fish. The torpedo-shaped body, large eyes and distinctive underbite are sure giveaways.

*Quote from *Live and Let Die* © Ian Fleming Publications Ltd., 1954; used with permission.

waters their seemingly random meanderings are deceptive, for in "passing gear" a barracuda can accelerate to 40 km per hour. And because of its impressive dentition, in parts of Australia the barracuda is known as a *dingo*, the Aussie word for a type of feral dog.

Though we're meant to fear for the demise of Agent 007 in Fleming's *Live and Let Die* underwater frightscape that opens this chapter, the savvy snorkeler or diver recognizes the plot line as more fairy tale than reality. In and around coral reefs, barracudas have little reason to find a human an appealing victim, unless the fish is drawn — often in murky conditions — to a glittering piece of jewelry or the reflection from an eager diver's camera lens. Also, a pack of large barracuda is generally a rare sight, as most tend to be solitary hunters. Juveniles, though, sometimes create beautiful spiral formations, a fishy "tornado" that may confuse would-be predators.

The Latin name for the Great Barracuda is *Sphyraena barracuda.* Some scholars believe *Sphyraena* derives from *sphyra*, meaning "hammer." *Sphyraena* first appears in 350 BC on a fish-laden list cobbled together by Aristotle, though how the barracuda resembles a hammer is one of those etymological briar patches perhaps best left unexplored. Greeks also referred to the barracuda as a *cestra*, a javelin used in the Persian war of the 5th century BC.

There are 20 species or so of barracudas. The largest is the Great Barracuda, once called *parricoota* or *picuda* in the West Indies, edging us closer to *barracuda,* the common name tossed around today. Formidable in appearance, barracudas seem to take a macabre pleasure in flashing a toothy grin that, in the diver's imagination, only seems to widen as the distance between scales and

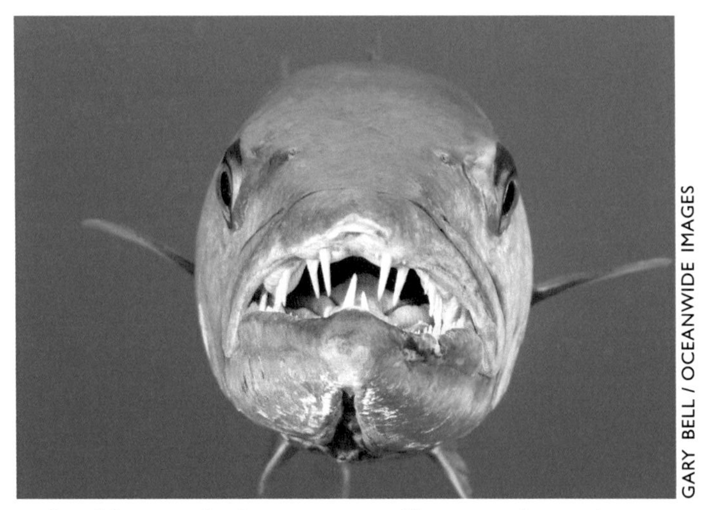

GARY BELL / OCEANWIDE IMAGES

A "smile" not to be forgotten: An efficient predator, the Great Barracuda uses long daggerlike teeth to secure its victim. A second set outside the first are used to tear the flesh of the prey.

MELVINLEE / DREAMSTIME

A "tornado" spiral of juvenile barracuda

skin shortens. For the uninitiated as well as the experienced reef explorer, the companionship of a Great Barracuda — and they commonly exceed a meter and a half — can be humbling.

Blood-curdling reports that stretch as far back as the 17th century have tagged the barracuda unfairly as a pugnacious predator on the hunt. Take this example from the writings of Charles de Rochefort, a French chaplain based in the Caribbean.

> *"Among the ravenous Monsters that are greedy of man's flesh...the Bécune is one of the most dreadful.... He lives by prey, and furiously fastens like a Blood-Hound on the men he perceives in the water: He carries away whatever he once fastens on, and his teeth are so venomous, that the least touch of them becomes mortal if some sovereign remedy be not immediately apply'd..."*
>
> — Charles de Rochefort, 1658, *The History of the Caribby-Islands*

Half a century later in 1707, writing in his *Natural History of Jamaica,* the Irish physician and part-time naturalist Sir Hans Sloane described the Great Barracuda as "very voracious," relying, no doubt, on the accounts of fishermen and the divers who collected his corals and sea urchins. Barracuda and other fishes were also bought at markets. Techniques of preservation relied upon methods prescribed in the English apothecary and collector James Petiver's "Brief Directions for the Easie Making, and Preserving Collections of all Natural Curiosities," excerpted at the right.

A contemporary of Sloane's — the French missionary Jean-Baptiste Labat (or simply Père Labat), a keen

JAMES PETIVER'S "BRIEF DIRECTIONS..."

ENGRAVER UNKNOWN / ALCHETRON

One of the greatest challenges faced by those collecting tropical specimens in the 18th century was how to properly preserve the coral reef fishes and invertebrates. Petiver's directions follow:

> *All small Animals, as Beasts, Birds, Serpents, Lizards, Fishes, and other Fleshy Bodies capable of corruption are certainly preserved in Rack, Rum, Brandy or any other Spirit; but where these are not easily to be had, a strong Pickle, or Brine of Sea Water may serve; to every Gallon of which, put 3 or 4 Handfulls of common or Bay Salt, with a Spoonful or two of Allom powdered... and so sent them in any Pot, Bottle, Jarr... Cork'd and Rosin'd.... You may often find in the Stomachs of sharks, and other great Fish, which you catch at Sea, divers strange Animals not easily to be met with elsewhere; which pray look for, and preserve as above.*

Many of Petiver's own curiosities can be seen in the British Museum, where they may surprise and delight every bit as much as they first did upon delivery over three centuries ago to Petiver's apothecary on Aldersgate Street, London.

naturalist, adventurer and engineer — also recorded his observations of flora and fauna. While stationed on Martinique and Guadeloupe in the French West Indies, Labat noted in his memoirs that barracudas are more likely to attack an Englishman than a Frenchman. The former, he thought, being a heavy eater of meats and of a "beefy" form, contrasted with the more "delicate-bodied and daintier-feeding Frenchman." In Labat's world, the Englishman will "produce an exhalation of corpuscles whose odor is more penetrating, which scatter farther, and which strike more on the organs of these animals." Labat also believed that the region's Carib Indians could track an Englishman by smelling his tracks and they thought the flesh of an Englishman more appetizing than that of their cross-channel neighbors. Why, he implies, should this not also be true of predatory fishes?

But the danger from falling victim to a nip from a barracuda pales in comparison to the risk of eating one. In some parts of the world, they are a culinary delight, but in others a nicely grilled barracuda steak can send you to the emergency room or even to the grave.

Sloane and others came tantalizingly close to understanding the true cause of what we now know as *ciguatera poisoning*, a toxicity of fishes that can prove deadly to humans. "According to its feeding on venomous or non-venomous food," Sloane wrote, " 'tis wholesome or poysonous to those who eat it; 'tis also noxious in some Seasons of the Year… and innocent in others, I suppose according to its Nourishment, by which now and then, it acquires so much poison as to kill immediately."

Ciguatera was further documented in the South Pacific in 1774 on board Captain James Cook's HMS *Resolution*. The symptoms are enough to make one think carefully before dining on any fish other than tinned tuna. The crewman John Anderson wrote of "a flushing heat and violent pains in the face and head, with a giddiness and increase in weakness; also a pain, or as they expressed it, a burning heat in the mouth and throat."

The symptoms are classic for food poisoning, but as if that weren't enough, they may be accompanied by weird changes in sensation. Numbness, tingling and an odd thermal reversal in which hot objects feel cold and cold feel hot are typical and may continue for weeks and sometimes years. When Captain Cook had a run-in with the toxin after dining on a fish "with a large ugly head," his reaction proved odd: "Nor could I distinguish

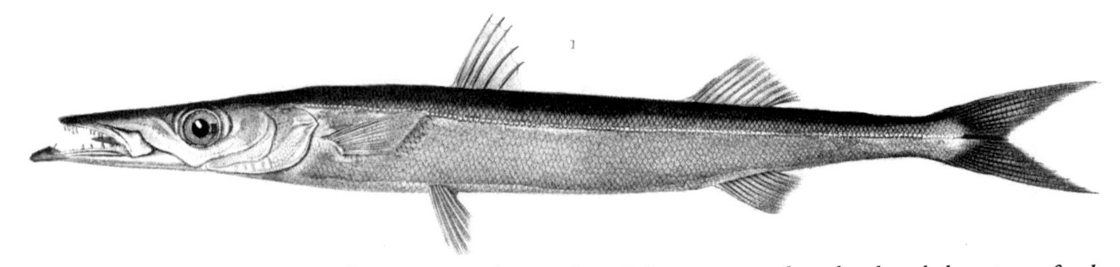

SOURCE: RAWPIXEL

Evidence of the collecting and documenting that continued into the 19th century, a hand-colored drawing of a barracuda graced the 1892 edition of Charles d'Orbigny's Dictionnaire Universel D'histoire Naturelle.

between light and heavy bodies, a quart pot full of Water and a feather being the same in my hand."

Ciguatera poisoning starts when a reef fish such as a parrotfish or tang dines on brown algae that has the toxic *Gambierdiscus toxicus* on the surface of its fronds. The toxin of this naturally occurring single-celled organism is not harmful to the fish that eats it, but is stored in its tissues and concentrated as it moves up the food chain. The trouble begins when the toxin reaches larger predators — sharks, mackerels, snappers, groupers and especially barracudas, at which point it can be very concentrated indeed.

So far, there is no effective test to detect ciguatera toxins in fish meat. Cultural "preventatives" for ciguatera poisoning include:

- Don't eat the fish if a silver coin cooked with it turns black.
- Avoid eating barracuda if the liver tastes bitter and peppery.
- If ants refuse to eat the fish, so should you.
- Fat fish are OK to eat, lean fish are not.
- Avoid fish with a bluish back.

Over time the barracuda has swum its way into numerous cultural avenues far distant from the coral reef. In 1964 a sports car, the Plymouth Barracuda, was introduced to America. An aerospace research firm, a jet, a boat, a bikini and an IT protection service company have attached themselves, remora-like, to the barracuda. The barracuda is a submarine and a fault zone in the North Atlantic. A vice presidential candidate of some notoriety once went by the moniker of "Sarah Barracuda." A bit of a row ensued when the signature tune of the rock group Heart, titled "Barracuda," was played at the 2008 Republican National Convention as a lead-in for Sarah Palin, and band members with a different political bent objected, proof still that no matter where a barracuda shows up, it can excite, upset and frighten — all at the same time.

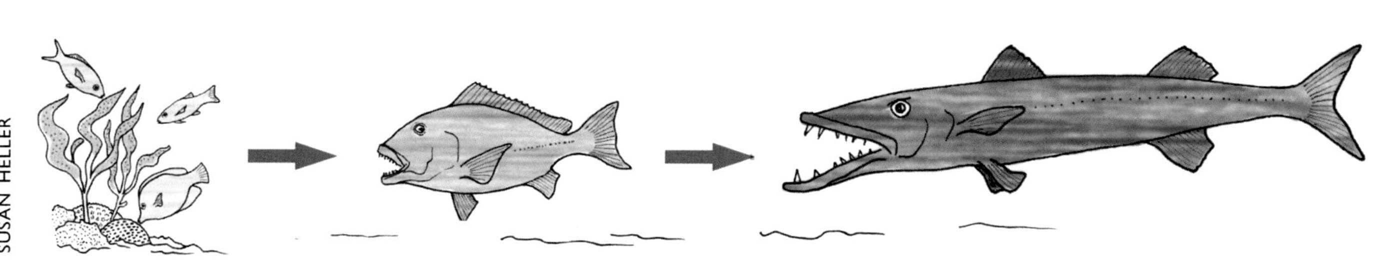
SUSAN HELLER

Harmless to fish, ciguatera toxins produced by the single-celled Gambierdiscus toxicus on brown algae are concentrated as they move through the food chain. The concentrations in higher-order predators can make their flesh lethal when eaten by humans.

15
The Perils of Exquisiteness

The turtle lays thousands of eggs without anyone knowing, but when the hen lays an egg, the whole country is informed.

— Malay proverb

If you are lucky enough to be standing at the right time on a beach edging the right coral reef, you might experience a bit of magic: a snout followed by an armored head that lingers at the surface and fits no preconceived notion of marine life. If you are luckier still and the creature is close to shore, then you may hear it breathe several times. A heartbeat or two later — though your heart may have skipped a few beats by this time — the water swallows the turtle and it may seem as if it was never there. As a snorkeler or scuba diver, to swim alongside or above one is as close to nirvana as we bipeds can get. In this dense medium that rewards sleek and streamlined swimmers with low energy costs, the sea turtle looks like the marine equivalent of a sumo wrestler that has lost its way — and one with which we could easily keep pace. Yet sea turtles are surprisingly swift and agile, capable of bursts of speed fast enough to elude sharks and killer whales, their only real predators other than man.

CAROLINE S. ROGERS

Few experiences match the thrill of seeing a Hawksbill Sea Turtle (Eretmochelys imbricata) as it soars across a palette of turquoise water.

The scientific name for the Hawskbill Sea Turtle pairs the Greek word for "rowing turtle," *Eretmochelys*, with the Latin word *imbricata* that describes the "overlapping scales," or *scutes,* so characteristic of the species. Hawksbills range the warm-water reef systems of Atlantic, Pacific and Indian oceans where coral reefs provide them with sheltering caves and ledges as well as a place to forage. Average-sized compared to most sea turtle species, they tip the scales at a svelte 70 kg.

These are carnivorous turtles, but don't look for them in ambush mode or expect to see them in hot pursuit of target prey. Though known to snack on the occasional anemone, sea cucumber or crab, the Hawksbill diet is mainly restricted to food sources shunned by just about every other sea creature: sponges and *zoanthids* (zo-AN-thids), tiny anemone-like animals, usually colonial. The Hawksbill uses its pointed beak to scrape its spongy prey from the reef's nooks and crannies. (That beak, by the way, is made of keratin, the same substance found in your hair and fingernails.) It's a diet that would land most animals in the emergency room, for not only are many of those sponges toxic, but some are laced with *spicules,* essentially minute shards of glass. Yum.

HUDSON FLEECE / ALAMY

A Hawksbill Sea Turtle feeding on a sponge at Tormentos, near Cozumel, Mexico, as two French Angelfish look on

Science has been slow to catch on to the mysteries of all seven sea turtle species, and gaps still exist in a basic understanding of their life histories. But we do know, for example, that a sea turtle's gender is dictated not by sex chromosomes X and Y — which it lacks — but by nest temperature during the middle third of incubation of the egg. Low temperatures are more likely to result in male hatchlings and high temperatures in females. Average nest temperatures produce a mix between the sexes, a balance that research has shown is now skewing female as a result of the warming of the planet. A recent study found that 99 percent of Green Sea Turtles hatching in the northern parts of the Great Barrier Reef are female.

Whether female or male, sea turtles have perfected navigational techniques far more sophisticated than the GPS systems you and I take for granted. They bring the turtle across thousands of watery, featureless kilometers to breed and lay eggs on the same stretch of sand where it was born. Ingenious and mysterious, the navigational system probably includes some combination of sun positioning, smell and an ability to sense Earth's magnetic field. It is an impressive bit of "technology" that brings the turtle home from distances up to 2,400 km away.

On hatching, a baby Hawksbill Sea Turtle snags a lottery ticket with just one chance in a thousand to win the ultimate prize — a life. Bubbling up out of the sand with its siblings, the hatchling makes a free-for-all dash to the

KLEBERPICUI / DEPOSITPHOTOS

On hatching, a sea turtle's first challenge is to make it to the water, where its longevity is anything but guaranteed. Here newly emerged hatchlings, still covered with sand from digging out of their nest, head for the water.

sea, its route patrolled by predators variously armed with claws, beaks and teeth. Now aquatic (if it survives), its fate is hitched to the currents, its longevity uncertain. Hiding out in a floating "mobile home" of *Sargassum* seaweed for several years is not such a bad idea.

There she nibbles on surface organisms and bits of seaweed, and strives to avoid entry into the food chain. Only when she is about dinner-plate size will our turtle venture back to the coast, where a new diet and now camera-toting tourists await. For millions of years, it was a strategy that worked: Hatch, scramble for the sea, grow larger, move to the coast, live for a long time, produce little turtles — a good life. And then one day, a sea turtle off Caribbean shores surfaced for air only to see an odd shape propelled by billowy white clouds.

When his ships were delayed off Cuba, Columbus blamed it on a "sea covered in turtles." Later navigators

CAYMAN ISLANDS POSTAL SERVICE

Cayman Islands quincentennial stamp issued in 2003 commemorating Christopher Columbus's voyage to the islands in 1503

sailing to the Cayman Islands from June to September could steer their way merely by listening to the sound of Green Sea Turtles heading there to lay eggs. Once sailors discovered that turtles, usually Green Sea Turtles, were an easily acquired source of meat, turtle steak and stew became the order of the day. The numbers of Green and other sea turtle populations in the Caribbean are probably less than 5 percent of the hundreds of millions that were there pre–European contact. Not so much a food supply — meat tinged with toxic sponges does not make for an appetizing meal, Hawksbill numbers tanked due to something much more insidious and far more global in reach.

Spotting a sea turtle never fails to thrill even the most jaded of divers, especially so the Hawksbill, named for its iconic hawklike beak. On a sunny day the Hawksbill's shell flashes intricate, luminous patterns in brown, orange and gold. As the turtle forages or rests on the reef, the shell is good camouflage, allowing it to blend in with its surroundings. Sadly, the shell's lovely sheen and patterning have led to its exploitation as a substance of value, joining elephant ivory, rhinoceros horn and any number of other natural "products" whose commercial value is exploited by black-market trading.

During the course of its life, a Hawksbill may travel more than 2,000 km between nesting and feeding grounds; in death, they travel the world. We don't know when the first artisan saw the value in a Hawksbill's shell segments, commonly known as tortoiseshell. In the 15th century BC, tortoiseshell was on Egyptian Queen Hatshepsut's shopping list during a trade expedition down the Red Sea. Ever since, it has become what the ecologist and science communicator Carl Safina, in *Voyage of the Turtle…In Pursuit of the Earth's Last Dinosaur,* calls "the world's oldest luxury good and quite possibly the most important motivator in 'globalizing' trade." As far back as the early 17th century, some people recognized the problem of overharvest of Hawksbills and other sea turtles. In 1620, so alarmed were the members of the Bermuda Assembly (originally the single house of the Parliament of the British territory of Bermuda) by the indiscriminate harvest of "so excellent a fishe," that they authorized *An Act Agynst the Killinge of Ouer Young Tortoyses:*

> *…from hence forward noe manner of pson or psons… shall pseume to kill or cause to be killed in any Bay Sound or Harbor or any other place out to Sea… any young Tortoyses that are or shall not be Eighteen inches in the Breadth or Dyameter and that upon the penaltye for everye such offence of the forfeyture of fifteen pounds of Tobacco…*

Yet the lure of such a readily available product continued to prove irresistible for traders and artisans, and tortoiseshell soon glowed from furniture inlays, weapons, jewelry, eyeglass frames, bracelets and berets, cigarette cases, model ships and much more.

SOURCE: METROPOLITAN MUSEUM OF ART

Left: An 1880s mask made of tortoiseshell with a frigatebird, produced by Mabuiag Islanders in the Torres Strait

Below: A 19th-century wooden mandolin with tortoiseshell inlay, made by Pietro Tonelli of Naples, Italy

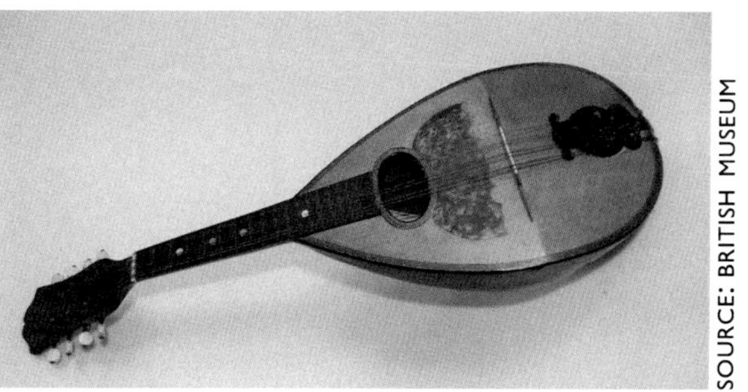

SOURCE: BRITISH MUSEUM

The statistics from the industry are jaw-dropping. During a two-decade period ending in 1992, scutes from 30,000 Hawksbills per year were imported to Japan for the *bekko* (tortoiseshell products) trade. In the '70s and '80s, a "fine souvenir" would have been a stuffed and polished juvenile Hawksbill. Over a half million of these sad mementos from Florida's Keys and beyond found their way into the tourist trade. Fortunately, the Hawksbill's rapid descent into oblivion was slowed in 1977 when most nations agreed to halt their exploitation. Degradation of habitat, poaching, and death by plastic and net continues, and by international standards Hawksbills still teeter at the edge of extinction.

Their demise would be catastrophic on several fronts, not the least of which is cultural. For millennia, all sea turtles have been a Jack-of-all-trades talisman across numerous lands. They have provided inspiration for stone carvers, coin designers and ceramicists. Mayan gods were adorned with the carapaces of sea turtles. Every four billion years or so, the Hindu god Vishnu becomes a turtle and goes about the business of renewing the universe. If you aspire to a long life, prosperity or fertility, or would like to hold hands with a guiding spirit in the hereafter, you might keep the sea turtle in your prayers.

Across the Pacific, sea turtles belong to the gods, or are gods themselves. They rescue fishermen, shift islands about on their backs, and shape landforms by rummaging about deep within the earth. Throughout Polynesia, dining on sea turtle was almost entirely restricted to chiefly and elite classes, taboos which may have helped keep turtle populations intact across thousands of years.

MULTI-TASKING

In fable and myth, sea turtles teach lessons that resonate even today.

One day, as a fisherman from Palau prepared to anchor his canoe over the reef, he saw an enormous Hawksbill Sea Turtle surface nearby. The man knew such a turtle would bring him fame and fortune. Neglecting to anchor his canoe, he dove into the water and after a long struggle, surfaced with the turtle. But when he looked about for his canoe, it had drifted away toward the horizon. The more he tried to swim the heavy turtle to his boat, the farther away the vessel drifted. When he finally gave up and freed the turtle, it was too late — his canoe had been lost and he had to swim back to his village, humiliated and with no fish, no turtle and now no canoe. The moral of the story:

Those who try to do two things at once often accomplish neither.

TARA BONVILLAIN

16
The Long Arms of the Maw

In Herman Melville's classic novel *Moby Dick* there is a point at which Captain Ahab stands on the bowsprit of the *Pequod* and spots what he believes to be the spume of the great white whale. He gives the order to lower the whaleboats and sets off in pursuit of his quarry.

They come upon the creature: "... a vast pulpy mass, furlongs in length and breadth, of a glancing cream-colour, lay floating on the water, innumerable long arms radiating from its centre, and curling and twisting like a nest of anacondas, as if blindly to clutch at any hapless object within reach. No perceptible face or front did it have; no conceivable token of either sensation or instinct; but undulated there on the billows, an unearthly, formless, chance-like apparition of life."

Silently the boats retreated to the *Pequod*. We rely upon the first mate, Starbuck, for an insight into the creature they've just met.

> *"Almost rather had I seen Moby Dick and fought him, than to have seen thee, thou white ghost!"*
>
> *"What was it, Sir?" said Flask.*
>
> *"The great live squid, which, they say, few whale-ships ever beheld, and returned to their ports to tell of it."*

Melville didn't just harpoon the image out of thin air, for such nightmarish monsters have been known for some time. He could have used *kraken* to describe it, a Norwegian word that first appeared in 1755 and meant "sea monster." The label was memorialized by Alfred, Lord Tennyson in the English Poet Laureate's "The Kraken," a poem best read aloud in whatever bone-chilling voice the reader can muster.

> *Below the thunders of the upper deep;*
> *Far, far beneath in the abysmal sea,*
> *His ancient, dreamless, uninvaded sleep*
> *The Kraken sleepeth...*

SOURCE: THE MARINERS MUSEUM AND PARK

Detail of an illustration from Sea and Land: An Illustrated History of the Wonderful and Curious Nature Existing Before and Since the Deluge *by James W. Buel, 1887*

Today we know the kraken of yore as the Giant squids (one to eight species, depending on whom you ask) or the Colossal Squid, among the largest animals ever to have lived. Just try to imagine what it might be like snorkeling above a coral reef when an 18-meter, 900-kg tentacled terror with soccer ball–sized eyes looms before your mask. Fortunately for snorkelers and divers, the kraken's habitat of choice generally lies at depths exceeding 500 meters. Fish watchers are much more likely to spot, rather than a kraken, a school of Caribbean Reef Squid or the tiniest known squid species, the Grass Squid, an eye strainer at a couple of centimeters long. It has been said that if you were to invent an animal, you would never invent a squid. The marine artist and author Richard Ellis describes them as "not part of our world, not elements of our consciousness.... Their

CAROLINE S. ROGERS

A school of Caribbean Reef Squid (Sepioteuthis sepioidea)

unfamiliar shape, with a cluster of arms at one end, eyes in the middle, and a tail at the other end, has only added to the impression that they are alien creatures from an unknown world — which is exactly what they are."

Like octopuses, squids are cephalopods, a Greek word that means "head-footed." Just look at a squid and the word makes sense. It has arms that seem to leap from its head. Melville could have painted an even more frightening word picture for his sea monster since most squid are also armed with two long tentacles used to capture prey. The added appendages jump squids into the category of *decapods*, two appendages up on the octopods. The eight arms, like those of octopods, and the two tentacles come equipped with suckers to grasp their prey. And the Colossal Squid takes nasty to an entirely different level — each of its suckers sports a scythe-like claw.

In the Coral Reef Olympics, squids might do quite well in the decathlon. They have a giant nerve fiber, an *axon* — one of the thickest in the animal kingdom — that gives them the ability to send messages to muscles much more quickly than most creatures.

Not only can they maneuver with astonishing speed, but they are also chameleon-like in their ability to change color. Unable to produce sound, squids contribute nothing to the underwater cacophony on the reef, but there is speculation they use their color-changing talents as a form of communication, an underwater semaphore system if you will.

A squid's skin, like an octopus's, contains *chromatophores*, cells it can control to produce a fantastical variety of spots and stripes. At the snap of a tentacle a squid can change hue, rotating through red, orange, yellow, brown

CAROLINE S. ROGERS

A Caribbean Reef Squid, hovering over a sandy bottom with its long feeding tentacles extended, looks straight at us.

CAROLINE S. ROGERS

Side view of a Caribban Reef Squid. Though they are mollusks, relatives of clams and snails, squids lack a shell.

PAUL SPRINGETT A / ALAMY

Striped, for now, the colorful Big Fin Reef Squid (Sepioteuthis lessoniana) is a quick-change artist.

and black. If neither camouflage nor flight works as a defensive strategy, a squid can try to bluff the pursuer with rapid changes in body patterns and color, by taking on a striped or spotted appearance, or even, like octopuses, elevating nodules called *papillae* on its skin.

A further technique is to release a "squidscreen" that allows the animal to slip away under the cover of a cloud of ink. Or, it might release a "pseudomorph," an inky replica of itself that confuses a predator, as well as any snorkeler who comes too close. But many creatures — whales, dolphins, fishes, seabirds and humans — have developed strategies for catching the elusive squid, and it has become an increasingly important fishery for diners worldwide who enjoy a meal of *calamar, lula, inktvis* or just plain squid.

The kraken that lolled on the surface as Ahab approached would have made for quite the "Nantucket Sleigh Ride" had the crew put a spear into it, and the men would have dined well that day. But let's take a different flight of fancy, twisting literary history — and cheating time while we're at it. Picture, if you will, John Steinbeck's *The Log from the Sea of Cortez* knocking about in the captain's cabin on the *Pequod.* Try to imagine what thoughts might have floated through Ahab's mind as he came upon this passage late that night, having run up on the giant squid earlier and at first mistaking it for Moby Dick:

> *Men really need sea-monsters in their personal oceans.... For the ocean, deep and black in the depths, is like the low dark levels of our minds in which the dream symbols incubate and sometimes rise up to sight like the Old Man of the Sea.... An ocean without its unnamed monsters would be like a completely dreamless sleep.*
>
> — from *The Log from the Sea of Cortez,* John Steinbeck, 1951

17

"A Marveilous Straunge Fishe"

As the 16th century unfolded and European explorers began their great voyages of discovery to the New World, big sharks were unknown in Europe, for the simple fact that large vessels had not ranged all that far out to sea. Spaniards were the first to come upon sharks of any size in this new frontier, soon followed by the English, whose previous experience with sharks had been limited to smaller species found around the British Isles.

As far as we know, the word *shark* was first used in June of 1567 during a fishing trip by the British Captain John Hawkins. His crew captured a shark of impressive size and hauled it back to London.

Unlike the fish, the word was not merely plucked from out of the blue. Some etymologists believe it derives from the Mayan word for shark, *xoc,* a word whose pronunciation — *shaak* — lends the notion a degree of plausibility.

Hawkins's shark was butchered on the docks, where some enterprising soul then had it stuffed and hauled off to a pub. It's not too difficult to picture what must have been the 16th-century version of an image and story gone viral, with long lines stretching the cobbled road of Fleet Street, curious eyes straining for a view of the bizarre creature suddenly landed in their backyard watering hole, the Red Lyon. And what a fish it was — by all accounts a 17-foot-long thresher shark! A handbill advertising the exhibition spread the word, proclaiming:

> *Here hath never the lyke of it ben seene. . . . Ther is not proper name for it that I knowe, but that certain men of Captayne Hawkinses doth call it a sharke. And it is to bee seene in London, at the Red Lyon, in Fleetstreete.*

Detail from a Thomas Colwell broadsheet advertising Captain John Hawkins's shark, 1567

No doubt, the eye-popping "straunge" fish thrummed the same primal chord many of us try to ignore when presented with a creature that appears quite capable of snipping us in two. It seems only natural that sharks are so wrapped up with words evocative of danger: Sharks *congregate, patrol* and *circle;* they *lurk* and *prowl* (*unpityingly* so); and eventually they *devour* with *razor-sharp* teeth. Sharks never *school,* they *infest* and always in a *frenzied* manner. But talk to any experienced divers who have been with and observed their share of sharks and those labels fall by the wayside, replaced by *fluid, graceful, sleek, sinuous, beautiful, cautious* and simply *fascinating.*

Why so fascinating? Because of their unusual skin and remarkable senses. The sandpaperlike surface of their skin, made up of tiny, overlapping toothlike structures called *denticles*, helps reduce the drag of the water as the shark moves through it. And lucky is the snorkeler or diver who gets to witness a shark shifting into "passing gear," best described as watching an underwater missile take flight at upwards of 65 km per hour. Whether the shark is casually swimming or streaking by, its senses are on full alert; its ability to hear low-pitched frequencies helps it detect injured or weakened fish. Having heard its prey from far away, a shark can use its electro-sensing ampullae of Lorenzini (see page 34) to home in on it.

FRANTISEKHOJDYSZ / SHUTTERSTOCK

Few experiences can match the thrill of going eye to eye with one or more sharks above a beautiful coral reef.

Unfortunately, we humans are hard-wired to "thrill" to the accounts of blood-thirsty predators of any stripe. Nor can writers resist the imagery of such a toothy animal:

> *For them the Ceylon diver held his breath*
> *And went all naked to the hungry shark*
>
> — from "Isabella," by John Keats, 1818

For William Shakespeare fans, *Macbeth* is a joy to read (and see) if only for its roller coaster ride of lush witchy language in Act IV, Scene I. Before a bubbling cauldron, a witch adds a snippet of this, a snippet of that, including our subject fish:

> *... Scale of dragon, tooth of wolf,*
> *Witches' mummy, maw and gulf*
> *Of the ravin'd salt-sea shark,*
> *Root of hemlock digg'd i' the dark ...*
>
> — from *Macbeth*, by William Shakespeare, probably written between 1599 and 1606

Culturally distant and centuries removed from Shakespeare's "Double, double toil and trouble ...," the fanciful phrase "jump the shark" refers to the point at which something no longer improves and only gets worse. It comes from a scene during the last years of the 1970s TV series *Happy Days,* during which the character known as "the Fonz" launches above a shark while water skiing. Critics said that from that point on, the show listed heavily to port. You could say Fonzie was doing his best to end up as shark bait, a solitary swimmer well out from shore. If nothing else, the Fonz was an honest sort, not *sharkish* — not a cheat. Nor was he ever known to *shark upon* anyone — to victimize or swindle someone.

Sadly, the bulk of news accounts in print and breathlessly recounted on TV are devoted to sharks attacking humans, when the opposite is the real story. Typically the score goes something like this:

Sharks killed by humans every year = 100 million
Humans killed by sharks = a dozen or so

Residents of the United States are 20 times more likely to be killed by a cow than bitten by a shark. Beware of those Holsteins.

It seems as if we are the ones who have taken a severe bite out of sharks. In the past three decades shark populations have fallen precipitously. According to the International Union for Conservation of Nature (IUCN), more than half of open-ocean (pelagic) shark species are at risk of extinction, with most of that decline due to overfishing for shark-fin soup — still a prestigious offering in many parts of Asia.

But the shark has a long swim to go to escape its bad reputation. One less shark in the sea doesn't ring the same sympathetic chimes as a beached whale does, or an abandoned seal pup, or for that matter any other creature lacking daggers for teeth.

LANO LAN / SHUTTERSTOCK

Shark fins drying in the sun, still an all too common sight

Many of the large sharks implicated in attacks — Tiger, Bull and White, for example — are rarely seen on the coral reef. Much more likely to be seen are the White-tips and Black-tips. They patrol reef fringes, watching and waiting for prey, promoting the image of the erratic, unpredictable, twitchy carnivore, when in reality they are of little threat to humans. Even more easily observed are Nurse Sharks — placid, benign dwellers of the shallows that often seem to be taking cat naps on the sea bed. The scientific name for the Nurse Shark, *Ginglymostoma cirratum,* sounds like part of a witch's curse and marries the Greek and Latin to mean "curled, hinged mouth."

The common name Nurse Shark might be related to *hurse*, an early English name for the dogfish, distant kin to the Nurse Shark. The *n* may have eventually supplanted the *h,* a practice not all that uncommon in medieval times. At some point in the 16th century, the word *nurse* was used to describe any large fish, but especially a shark. The name came unstuck, of course, when Captain Hawkins deposited his beast on the London dock and *nurse* became *shark.* (Use your imagination, dear reader, and envision a watery world where fishes called nurses and not sharks glide silently past. A bit easier to hop into the water, isn't it?)

And what could be more of a treat than to learn of the benevolent side of what surfers are fond of calling "the men in gray suits." In one Polynesian myth, Kauhuhu, Shark God of the Hawaiian island of Moloka'i, was known to devour all men who dared enter his cave. One day the mortal, Kamalo, risks a visit to Kauhuhu, seeking revenge for the murder of his two sons by an evil chieftain. Kamalo manages to tell Kauhuhu his story without becoming that day's main course. Taking pity on him, the shark god agrees to help and sends a terrible storm that carries the evil chieftain and his followers into the bay, where they are devoured by... well, guess what. In payment, Kauhuhu claims the beautiful Lanilani, Kamalo's daughter, as his mate, changing her into a tiger shark. It is said Lanilani still watches over the people of Moloka'i and can be seen rising to the water's surface to enjoy the sound of the great hula drums played during island festivals.

Elsewhere, Taputapua is a Polynesian shark deity who provides a sort of inter-island transport for the ancestors, useful in times of family crisis or squabbles. Since Taputapua frequently takes vengeance upon wrongdoers, spear-fishermen — who already tempt sharks by towing bloody fish through the water — are often loath to go diving if they are quarreling with wives or other family members. If given a vote on the matter, reef fishes, including sharks, would probably like to see more such bickering.

GARY E. DAVIS

Harmless unless provoked, a Nurse Shark (Ginglymostoma cirratum) and the remora (suckerfish) that travels with it rest on the seabed Exuma Cays Land & Sea Park, Bahamas.

HUBERT YANN / ALAMY

Solo or schooling, sharks like this Grey Reef Shark (Carcharhinus amblyrhynchos) swimming over a coral reef are a magnificent sight.

18
A Reef Fish That Does Headstands

You might not recognize a trumpetfish when you first spot one. It makes no sense whatsoever that it should even exist: a fish shaped like a flute, with a head like a horse's connected to its teeny little tail by a long tube of a body up to about 60 cm in length. Plenty of coral reef fishes have advanced along their evolutionary track with poorly designed locomotive abilities — think trunkfish, frogfish, even the seahorse. But none seem to have been so tragically kitted out for negotiating their marine world as the trumpetfish. The poor chap's only means of compensating for such a challenging design seems to be a mad churning of pectoral fins to move forward or in reverse. Swimming is further aided with an energy-saving technique known as *balistiform* propulsion, involving simultaneous undulations of the dorsal and anal fins. As if exhausted by such efforts, our friend will as often as not tilt up end over end, seeming to cast its fate to the undulating whims of the water current.

Trumpetfishes reach a maximum size of about a meter, much of which seems to be taken up by the mouth, an impressive piece of architecture capable of ingesting fish larger in diameter than the trumpetfish's own body. The family name, Aulostomidae (oh-luh-STOM-ih-dee), comes from the Greek *aulos*, which means flute, and *stoma*, meaning mouth.

GARY E. DAVIS

Standing on its head, a West Atlantic Trumpetfish (Aulostomus maculatus) is a master of blending in, changing position or color to match its surroundings.

There's just one genus of trumpetfishes, three species. But a reunion of the order Syngnathiformes (sing-nath-ih–FORM-eez), to which trumpetfishes belong, would be reminiscent of the Star Wars bar scene — a gaggle of outlandish relations indeed, including cornetfishes, seahorses, pipefishes and flutemouths, elbow to elbow with flying gurnards and seamoths.

"Going vertical" is just one of trumpetfishes' stealthy fade-into-the-background tricks, a way to see and not be seen, especially if tucked within a stationary hideaway such as the frilly plumes of a lush *gorgonian* (soft coral). Trumpetfishes rank right up there with squids and octopuses in their ability to assume the correct hue to suit their cover, and the jet-blue tint of a trumpetfish in a gathering of blue tangs is a dazzling show to behold.

JOHN ANDERSON / DREAMSTIME

A West Atlantic Trumpetfish (Aulostomus maculatus) hangs out in a school of Atlantic Blue Tang (Acanthurus coeruleus) on a reef in the Caribbean Sea.

In appearance and movement, the trumpetfishes are the coral reef's version of an attack submarine. And when a trumpetfish pretends to have enrolled in a school of tangs, or slips alongside or above a parrotfish or grouper, it achieves a pinnacle of craftiness. As if the trumpetfish were looking to establish friendship at the interspecies level, this strategy allows it to edge within inhaling range of unsuspecting prey such as small fishes, shrimp and crabs.

Aulostomus maculatus was first described by the French zoologist Achille Valenciennes in 1837. Valenciennes was a protegé of Georges Cuvier and worked with him on what became a dense, never completed reference, the *Histoire Naturalle des Poissons.* A compilation of virtually all that was known about fishes up to that time,

JOHN ANDERSON / DEPOSITPHOTOS

Using a strategy known as shadow feeding, a trumpetfish usually hovers near benign herbivores, hiding to get close to its prey. But here it shadows a Tiger Grouper (Mycteroperca tigris), an efficient predator, perhaps for protection.

its 11,000 pages spread across 22 volumes described an astounding 4,055 species. More than half of these were new to science. Cuvier's intention was to chronicle all of the planet's known fishes, but he died before the work was completed. Even by today's standards, its detailed illustrations leave one breathless.

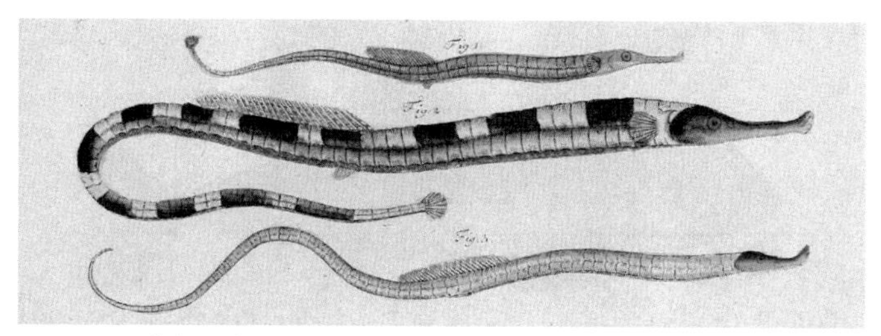

Trumpetfishes from Histoire Naturalle des Poissons

As a young man, Cuvier lived on the Normandy coast and discovered his niche while doing what many of us do at the beach — collecting and puzzling over the curiosities that wash ashore. The beachcomber begat the man, who became a full-time naturalist, and the seashore came to dominate his time and thought. The fishes were by all accounts his favorite animals. A specialist in ichthyology and comparative anatomy, by 1795 he was the era's "Google equivalent" when it came to fish, and naturalists around the world competed to aid him with notes, manuscripts, drawings and specimens.

One can only wonder if Cuvier ever stared into the waters, wishing for a way to swim with the creatures that consumed him so. It's an engaging thought: Imagine him hanging from the side of a boat, head beneath the surface, his senses alive to the reef's rampant pulsations. Aside from color and architecture, he would have marveled at the chorus of clicks, snaps and other sounds that surprise and delight so many reef visitors. He might have reconsidered his thinking about fish ears. According to his memoirs, he thought the construction of their ear "scarcely allows them to distinguish the most striking sounds; and, in fact an exquisite sense of hearing would be of very little use to those destined to live in the empire of silence…"

Cuvier could not have imagined that a submerged vessel might one day resemble his trumpetfish (and for that matter be rather keen on hearing what this liquid world has to say). The similarity between trumpetfishes and submarines did not go unnoticed by the United States Navy. In 1945, SS-425 — the USS *Trumpetfish* — slid into the water for the first time. (There were so many submarines to be named at the time, it must have been quite a challenge to come up with labels of an assertive, combative nature. *Barracuda, Thresher* and *Hammerhead* cruised the seas in those years. But so did *Bluegill, Carp, Catfish* — and *Trumpetfish.*) In her day, the sub patrolled the waters off Key West and in the Caribbean. So we can imagine *Trumpetfish* sliding silently past a trumpetfish in a bizarre underwater *pas de deux.*

A patch produced by BC Patch LLC commemorating the USS Trumpetfish *submarine*

19

You Laugh, Vex Chief, He Break Your Head with Club!

Who would have imagined that a little soft fish could have destroyed the great and savage shark?

— Charles Darwin, *The Voyage of the Beagle*

Spiny creatures that can reach a length of a meter, porcupinefishes are found in all tropical seas around coral reefs and mangroves, where they travel alone or in pairs. They can seem a tad blasé when venturing from the shelter of the reef, but look no further than the name as a clue to their insouciance. When threatened, the porcupinefish gives a whole new meaning to the term "hyperinflation." Capable of rapidly sucking in water (or air if a "fish out of water"), a perturbed porcupinefish can puff up to several times its original size. Its sharp spines stand out, transforming it into a prickly basketball no one would want to dribble. Predators are little interested in the spiny treat, and so it would seem the fish has — fins down — the perfect defense. Darwin thought as much, knowing of Houdini-like examples wherein a porcupinefish, swallowed live by a shark, had gnawed its way to freedom, redefining — at least for that shark — the term "upset stomach."

DAVID FLEETHAM

DAVID FLEETHAM

The flexible spine, elastic stomach and stretchable skin of the Spot-fin Porcupinefish (Diodon hystrix) allow it to go from unremarkable to extraordinary.

Such a talent requires an impressive set of choppers, and in 1758 Carl Linnaeus latched onto the teeth as a way to scientifically describe the Spot-fin Porcupinefish as *Diodon hystrix.* The Diodontidae (dye-oh-DON-tih-dee) are a family of fishes with two fused teeth (thus the name) per jaw. *Hystrix* comes from the ancient Greek for porcupine. With their powerful mashers, *D. hystrix* and other porcupinefishes are able to crush the hard shells of their prey — snails, sea urchins, hermit crabs — and while we're on the subject, the prodding human finger as well.

Porcupinefishes are closely related to pufferfishes, and to varying degrees all puffers have an effective passive defense, a deadly neurotoxin — *tetrodotoxin (teh-TRO-doh-toxin),* or TTX — concentrated in the ovaries, liver and skin. Puffers are even used by Haitian voodoo doctors to concoct a powder that is said to change people into zombies. With that as a recommendation, it's hard to imagine why anyone would think of making a meal of these fish. And yet, people do.

NICESCENE / SHUTTERSTOCK

A Tiger Puffer (Takifugu rupripes) on a Japanese postage stamp (c. 1966)

In Japan, dining on *Takifugu rubripes*, commonly known as the Japanese Puffer, Tiger Puffer or just *fugu*, is an adventure not to be taken lightly. Specially licensed chefs undergo rigorous training to learn how to safely remove the toxic portions of the fish. Figured into the equation before fugu is portioned out are a diner's weight and sex. Gourmands are well advised not to have seconds, since the risk goes up with increased consumption. And the attraction? A talented chef is able to escort fugu fanatics to the precipice of death, where a "satisfied" diner will experience the sublime feelings of tingly lips and blurred vision, essentially the early symptoms of *tetrodotoxication,* one of the most virulent forms of food poisoning. In 1991 the hazard was emphasized when TV personality Homer Simpson gave his fans a scare, narrowly escaping a case of fugu poisoning.

Though deaths have dropped off from a high of 176 in 1958, when chef licensing began, each year 10 to 20 people still pay more than they had planned for the culinary adventure. Or maybe some were jilted lovers following advice found in the 18th-century Japanese poet and painter Yosa Buson's verse, translated here:

I cannot see her tonight
I have to give her up
so I will eat fugu

One need not visit a coral reef to see a porcupinefish, as they can be found in souvenir shops, where their dried, shellacked and varnished skins are mounted on pedestals or even, perversely, made into lampshades. Even as far back as the early 18th century, the dried husks of porcupinefishes, boxfishes and other specimens were exchanged in payment for drink by financially strapped sailors. Many a London pub hung such mementos as items of decoration and attraction, where more than a few apparently caught the eye of the ichthyologist Peter Artedi (introduced on page 58), subsequently to find a home in his ichthyological opus.

But at least one group of Pacific Islanders employed the skins of porcupinefishes somewhat more creatively, as noted by Dr. John Coulter, a young 19th-century Irish physician with a healthy appetite for South Seas adventure. In 1832 Coulter began a four-year journey on board the English whaling vessel *Stratford*, attracted by the notion of exotic and strange South Seas experiences. When not tending to ailing sailors, Coulter had his share of adventures, which included lancing a whale and spending two weeks on a solo expedition in the Galapagos Islands. In the Marquesas he was tattooed, not out of vanity but to avoid what he viewed as the real possibility he might otherwise end up in the cooking pot! But it was in the Kingsmills (present day Gilbert Islands) where Coulter's travels briefly intersected with a porcupinefish.

This was a time of frequent intertribal conflict, when Gilbertese men were raised to be warriors, outfitted with a full "suit" of protective armor. "The head is surmounted by an extraordinary apology for a helmet... made of dried fishes skin," Coulter wrote. To fabricate each helmet, islanders captured a porcupinefish and once the fish had inflated, buried it in sand. After a week, the now hardened fish was exhumed and tailored to make a rather formidable spiked helmet, tied around the chin with a strand of woven coconut fiber.

Upon seeing a group of warriors thus outfitted, Coulter couldn't help himself and began to "shake with a concealed laugh." Wisely, an interpreter stifled his moment of merriment, and probably saved his life, saying "You laugh, vex chief, he break your head with club!"

SOURCE: BRITISH MUSEUM

A mid-19th-century porcupinefish helmet from Kiribati in Micronesia

20
Queer Fish

"Queer fish" is British slang for a strange person. But in Mabel Dwight's lithograph of the same name, it seems ambiguous as to which is the odder of the two — the fish or the man.

> *... One day I saw a huge Grouper fish and a fat man trying to outstare each other; it was a psychological moment. The fish's mouth was open and his telescopic eyes focused intently. The man, startled by the sudden apparition, hid his hat behind him and dropped his jaw, also; they hypnotized each other for a moment — then both swam away. Queer Fish!*
>
> — Mabel Dwight, unpublished autobiography begun in 1941*

With an unfishlike penchant for curiosity, groupers can seem a bit off. The reef's version of the nosy neighbor, they like to closely examine those entering their watery kingdom. Some groupers even partner with moray eels to go on hunting expeditions (see page 39), one of the strangest of the reef's many pairings. Queer fish indeed.

Not that a grouper needs much help finding food. It is one of the reef's top carnivores. Two-thirds fish and one-third dining machine, the grouper has a capacious mouth that harbors an impressive array of teeth, angled to the rear, that are used to catch and hold onto prey. There are some 150 species of grouper, ranging from the flyweight 2-kg Coney to the heavyweight Giant Grouper, up to 2.7 meters long and tipping scales at a prodigious 400 kg. With a stout body and a lower jaw that protrudes in a cocky manner, a grouper on the prowl looks like a boxer

SOURCE: ACHENBACH FOUNDATION FOR GRAPHIC ARTS

Queer Fish *(lithograph on stone) by Mabel Dwight, 1936*

* Quoted in *Mabel Dwight: A Catalogue Raisonné of the Lithographs* by Susan Barnes Robinson and John Pirog

CAROLINE S. ROGERS

Nassau Grouper (Epinephelus striatus) and Spotted Moray (Gymnothorax moringa) near St. John, U. S. Virgin Islands

scrapping for a fight. They are generally solitary, and when not hunkered down in their dark lairs, they skulk about the reef's shadowy alleys as if looking for trouble — which you, dear diver or snorkeler, seem to represent. But in fact they are anything but threatening, acting more like the reef's version of a curious puppy dog.

The name *grouper* comes from the Portuguese *garupa,* which was probably taken from a now extinct language once spoken by the Tupi tribe of Brazil. In the 16th century, Tupi was the language used by Europeans throughout Brazil to communicate with Amerindians. (The Tupians must have been even more impressed by the dentition of another carnivorous fish; the literal translation of the Tupi word *piranha* is "toothed fish.")

In 1519, the Tupi likely traded *garupa* for nails and knives with some ships of historical significance. Commanded by the explorer Ferdinand Magellan, the armada was searching for a sea route to the Spice Islands of Indonesia. Their rendezvous with history as the first maritime expedition to circumnavigate the world met with many setbacks after they left Brazil, most notably the loss of their commander in the Philippines. There, Magellan battled the forces of chieftain Lapu-Lapu, partly because the latter had no interest in converting to Christianity. Badly outnumbered, Magellan met a grisly end on the shores of Mactan Island, or perhaps it was Poro Island; historians differ. The home team's victory is commemorated in various ways. There is a 20-meter-high statue of Lapu-Lapu close to the harbor on Mactan Island; a city named Lapu-Lapu; and one can even dine on a meal of grouper — yes, better known locally as Lapu-Lapu.

When not on a dinner plate or searching for dinner themselves, groupers do what comes naturally — they make more groupers. But they go about it in a curious way. Many species of groupers are *protogynous hermaphrodites (pro-TAW-jin-us her-MAFF-row-dites),* a mouthful of an expression meaning they change sex — irreversibly so — from female to male. The "plumbing" as it were, of female groupers contains male gonads primed to express themselves. The trigger for this makeover is one of the mysteries of the coral reef, but for some species it appears to be linked to mass spawning events. As a preliminary to creating little groupers, adult groupers

OLGA NOSOVA / DREAMSTIME

Lauded as a hero in the Philippines, Lapu-Lapu was originally from Borneo.

participate in spawning aggregations that include social interactions — think singles-only dances — that give them a chance to assess the relative numbers of males and females. If there are too few guys in attendance, some females will spend the months before the next aggregation transitioning. The dances and socializing were all going just swimmingly for years, until one day fishermen began to crash the submerged spawning party. Case in point, the Nassau Grouper.

Once common throughout the Caribbean and West ern Atlantic, Nassaus are easily distinguished from their

IMAGES & STORIES / ALAMY

Nassau Grouper mass spawning event in January, 2007 at Glover's Reef, Belize

brethren by five striking vertical dark bars on a buff background. At maturity they may weigh upwards of 20 kg). Unfortunately, big Nassau groupers have suffered mightily for their sexual proclivities. When the mood strikes, as it does during full moons between December and March, large Nassaus will swim as far as 250 km to traditional offshore breeding grounds. Loyal to a fault, they return year after year to the same place in numbers that may reach 100,000 individuals.

Gatherings of such magnitude are irresistible for fishers, who make big catches, targeting the largest fish. In many locales, Nassau Groupers are no longer a part of the reef tapestry, having been fished out. Fifty years ago, Nassaus caught in the Florida Keys averaged about 18 kg; by 2007 that figure had dropped to just 2.5 kg. They are now protected in many places, but poaching continues and their long-term viability is tenuous.

Groupers play a significant role in maintaining the balance and diversity of a reef community. On the one hand, if unchecked by predators such as the grouper, the fishes and invertebrates that harm corals can increase and prove detrimental to the reef. On the other hand, when groupers eat parrotfishes, they remove species important for keeping certain kinds of algae in check. Groupers also snack on the occasional lionfish, and maintaining a robust grouper population may prove significant in controlling the spread of that non-native species throughout the Caribbean (the lionfish invasion is described on page 98).

A PASSION FOR FISHES

The Nassau and other groupers are just a few of the more than 1,400 fish linked to a single inquisitive 18th-century soul. Marcus Elieser Bloch, born in 1723 to a poor German family, became one of the leading ichthyologists of his time. Getting a late educational start (he was illiterate at age 19) didn't deter him in the least, and Bloch became a physician with a late-blooming passion for natural history. He began aggressively collecting animal specimens, especially fish, to keep in his Berlin apartment in a series of glass cupboards. (Among other curiosities, Bloch's cabinets held intestinal worms, some 400 stuffed birds and their nests, as well as embryos — mostly human.) Over half of the fish collection survives, one of the oldest in existence. By rights, it could be named for his wealthy second wife, Cheile, who funded the adventure, willingly or not, sacrificing sapphires for scales in the interest of science.

No other ichthyologist before Bloch had assembled work that so carefully described and depicted such a large number of fish species, many introduced to a broad audience in color and name for the first time. His work became a part of one of the most arresting publications on fish ever produced. The 12 volumes of his *Oeconomische Naturgeschichte der Fische Deutschlands* includes more than 400 illustrations, carefully hand painted by artists who even used silver-based paint to lend a lifelike sheen to fish scales.

For his time, Bloch may have seemed a queer fish himself, with an unbounded passion for finned creatures he would never see in the wild. Just envision the raised eyebrows of curious neighbors as crate after crate of pickled and dried specimens land on his doorstep. Let's enter the apartment now, where we see Bloch, in his lab, a container on the desk before him. It may be from Dr. Isert in the Lesser Antilles or Father John in Tranquebar, India or from one of his many other suppliers in far-flung places. He pries the lid off, removes the specimen. His eyes widen with disbelief as they take in the grouper or other fish, a creature new to him, new to science.

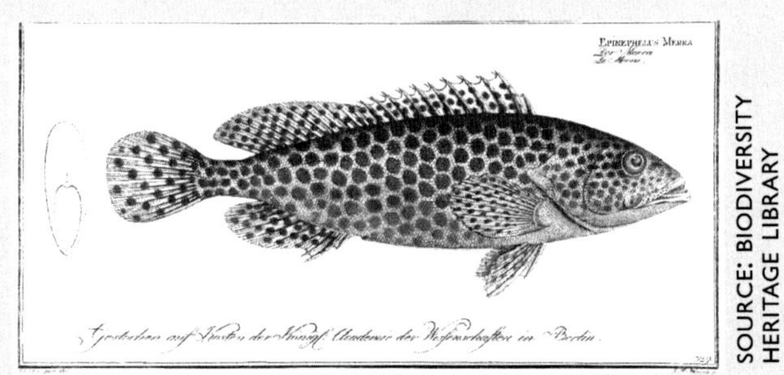

SOURCE: BIODIVERSITY HERITAGE LIBRARY

The Honeycomb Grouper (Epinephelus merra) from Marcus Elieser Bloch's Naturgeschichte der Fische

21
Super Males

The pressed carcass of a Longnose Parrotfish (Hipposcarus harid) from the "fish herbarium" at the Natural History Museum of Denmark

Tucked into the archives of the Natural History Museum of Denmark lies a fish, catalogue number P5952. Little more than a dry, headless, tattered skin, the creature is unremarkable but for the method in which it has been preserved for nearly two-and-a-half centuries. Rather than floating in a jar of formaldehyde, the fish has been pressed, plantlike, as if destined for an herbarium. The catalogue pegs it as *Hipposcarus harid* (Forsskål, 1775), commonly known as the Longnose Parrotfish.

Even on the most colorful of reefs, parrotfishes' flamboyant blends of greens and blues, highlighted with reds and yellows, make these stout-bodied fishes stand out, and when they school it can seem as if a fully decked-out parade band is marching by. They are abundant on coral reefs and some of the largest fishes to be seen there. One never forgets the experience of bumping into a Bumphead Parrotfish, a species that can exceed a meter in length, tipping the scales at a healthy 45 kg or more.

WATERFRAME / ALAMY

A school of Bumphead Parrotfish (Bolbometopon muricatum) parades over a coral reef in the Pacific Ocean near Indonesia.

The parrotfish family tree of 80-odd species is adorned with an assortment of beguiling names; if so inclined, one could parse out the defining characteristics of the Humphead, Bumphead, Bullethead and Globehead parrotfishes. Mastering those, there are still the Bicolor and Tricolor, Daisy and Dotted, the Bluemoon, the Stoplight and of course the Common Parrotfish. Little wonder that these alluring models glide across the faces of many an island-nation postage stamp.

The Caribbean nation of St. Kitts and Nevis has a separate postal authority for each of the two islands. In 2007 Nevis issued a block of four parrotfish postage stamps.

Of all the mysterious clicks, snaps, grunts and other sounds heard on the coral reef, one that's easy to identify is the crunch and munch emanating from the parrotfish café. Parrotfishes have teeth fused into two beaklike plates, an imposing jaw that resembles the bill of their avian namesake. So equipped, they seem to be nonstop diners, and a healthy coral reef will almost always include their patronage. They are responsible for "weeding" the reef of excess algae that if unchecked could easily smother corals. Indelicate diners, they are guilty of ingesting bits of the coralline structure along with the algae, and the scars of this activity are easily seen on coral heads. The bitten-off coral is ground up by specialized pharyngeal (fah-RINJ-ee-al) teeth in the back of the throat, then moved through the fish and jettisoned as fine sand over the reef. Parrotfish poop is one of several sources of the stunning white sandy beaches seen in the tropics.

DAVID FLEETHAM / ALAMY

A Bumphead Parrotfish deposits coral sand near Yap Island, Micronesia

Dining habits of parrotfishes are second in curiosity only to their sexual proclivities. Like certain species of groupers, wrasses and gobies, parrotfishes can change sex, with some females becoming fully functional males; animals that are born female and at some point in their life become males are referred to as *protogynous hermaphrodites,* as described on page 86. Populations that have the two types of males — original and protogynous — are called *diandrous,* meaning two-males. But there's still a third phase, a "terminal" protogynous male, larger and more colorful than all the others. Known as a "supermale," he defends a territory against other males and courts and fertilizes a harem of females protected within his realm. Should the supermale enter the food chain, the job opens up and another female switches gender and fills the position, a switch that may take as few as five days.

Some parrotfish species, when they need to rest from their exhausting social and feeding activities, can simply tuck themselves into a reef crevice for security and sleep the night away. As the reef darkens and the nocturnal shift of fishes and invertebrates comes on duty, a remarkable change occurs. In less than an hour, specialized glands under the parrotfish's gills secrete a mucous cocoon that soon envelops the fish in a gauzy "sleeping bag" of sorts. The theory is that the cocoon keeps ectoparasites at bay, acting almost like a submerged mosquito net.

The carcass of the Longnose Parrotfish at the beginning of our story is linked to a tale of adventure and perseverance not widely known or appreciated. On January 4, 1761, six men sailed on what was billed by the *Copenhagen Post* as a voyage of "discoveries and observations for the benefit of scholarship." They were off to study the nature

IMAGE COLLECTION / ALAMY

A supermale Rainbow Parrotfish (Scarus guacamaia) stops at a cleaning station for grooming. Rainbow Parrotfish can be more than a meter long and 20 kg in weight.

REINHARD DIRSCHERL / ALAMY

A protective "cocoon" envelops a sleeping Steephead Parrotfish (Chlorurus strongylocephalus) at Baa Atoll in the Republic of Maldives.

and culture of an unknown and exotic region, the southern tip of the Arabian Peninsula (now Yemen). The area was called Arabia Felix, loosely translated as "Happy Arabia" and regarded as nothing short of a verdant paradise.

Denmark was following the lead of other governments in pushing out into the unknown during this Age of Enlightenment. With its emphasis on sampling coral reef fauna as well as terrestrial, the Royal Danish Arabia Expedition venture was ahead of its time. One of the six men was Petrus Forsskål, a Swede who at 28 years old had positioned himself as a brilliant thinker and naturalist with a strong interest in coral reefs. His role was to collect as many floral and faunal specimens as he could lay his hands on, and he would do exactly that, describing 151

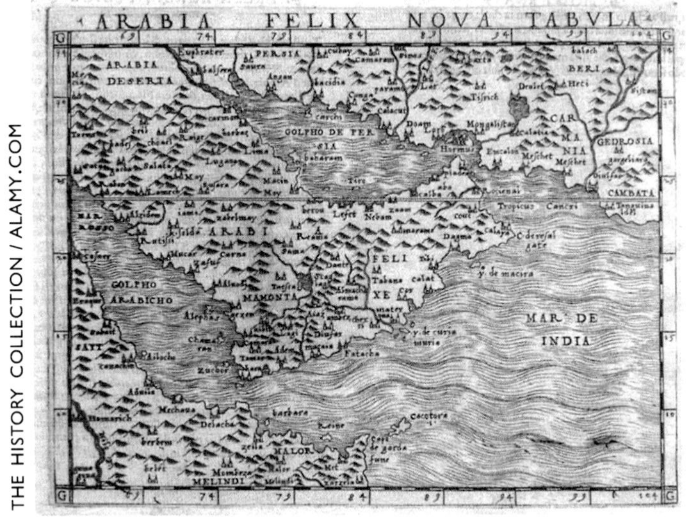

"Arabia Felix Nova Tabula," a map of the Arabian Peninsula credited to cartographer Giacomo Gastaldi and published in Venice, had been around for about 200 years when the Royal Danish Arabia Expedition was launched in 1761.

fish species new to science, including *Hipposcarus harid* (the Longnose Parrotfish shown at the start of this chapter) and several other parrotfishes.

Some of Forsskål's work was conducted around the Red Sea port of Jeddah, and one can only imagine the exotic distractions as he navigated a strange culture, bartering with fishermen for specimens, the atmosphere heady with the competing fragrances of coffee, frankincense, myrrh and cinnamon. In his journal, he wrote: "The people, the country, nature, everything was new to me; all the plants were new. I could think of nothing else but collecting and observing." But sadly, the price extracted for such knowledge was calamitous, and one by one — Forsskål included, the team succumbed to disease. Happy Arabia this was not.

Six years after setting off, the lone survivor, mathematician and cartographer Carsten Niebuhr, made it back to Copenhagen, where he made certain that Forsskål's work was not in vain (though countless specimens were lost or damaged in transit). Thanks to Niebuhr, the Longnose Parrotfish and others in Forsskål's Fish Herbarium, are still used today as species references.

Forsskål died having left a mark on science like few of his contemporaries. He had found unbounded satisfaction in making countless discoveries new to science, his contributions still recognized as remarkable for the time. One would like to believe that as he moved in and out of his malaria-induced coma those last days, the Longnose Parrotfish and other new-to-science reef creatures swam before his eyes, and that he died contented, having found true happiness in Arabia Felix.

22
Surgeons of the Sea

The scientific name for surgeonfishes, Acanthuridae (uh-can-THUR-ih-dee), comes from the Greek *acantha*, meaning thorn and *oura*, which means tail. The name is an appropriate one since these "surgeons of the sea" have scalpel-like spines on both sides of the caudal peduncle, where the body narrows before the tail. The spines are sharp, and if threatened or disturbed by a troublesome neighbor, scuba diver or fisherman, the surgeonfish deploys the scalpels and slashes at the intruder. The injured party, smarting from the painful wound, will think twice about a repeat engagement.

MATTHEW BANKS / ALAMY

A school of Powder-blue Surgeonfish (Acanthurus leucosternon) feasts on algae in the Andaman Sea near Thailand

The extended surgeonfish family includes tangs and unicornfishes; all have compressed, flattened bodies that look as if someone has taken a rolling pin to them. They are some of the most abundant and colorful of tropical marine creatures, favorites of many a coral reef visitor, including poet and avid scuba diver, Ingrid Wendt. One such species became the binding theme in Wendt's book of poems *Surgeonfish*. In the title poem, she tells the story of a painful attack and then describes the beautiful colors:

. . . and the surgeonfish

two of them, named for the bright orange, scalpel-sharp
fins at the base of the tail, striking
from out of the blue, head on, then swerving:

flak glancing the whole

length of our bodies, again and again. We must have looked funny,
flailing, thrashing at the surface, like runners dodging a sniper,
like puppets unstrung.

(continues on page 94)

But they were beautiful. That's
what we'd come for (such color!) the neon
indigo stripes of the black-scaled surgeonfish, neon.

indigo edging the top and bottom black fins edging the tail;
that pectoral fin, close to the gills: such a bright daffodil
sun! And this just one of so many others named in my book…

— from the poem "Surgeonfish" by Ingrid Wendt*

Most of the approximately 80 species of surgeonfishes are found in the Indo-Pacific region, and most are on the small side, no more than a foot or so long. But there's always an outlier to skew the average, and in the case of surgeonfishes, that's the Whitemargin Unicornfish *(Naso annulatus)*, the sumo wrestler of the family.

If there were a coral reef comedy club, a unicornfish would be the headliner. Snorkelers have been known to giggle uncontrollably just looking at the absurd "horn" jutting from the forehead. The exact function of the appendage? No one seems to know. Many unicornfish species start out as grazers, but eventually that pesky spike gets in the way and as adults they prefer dining on the microscopic free-swimming animals called *zooplankton.*

HENNER DAMKE / DREAMSTIME

The Sohal Surgeonfish (Acanthurus sohal), native to the Red Sea, has a reputation for slashing with the orange "blade" on its flank when it perceives a need for self-defense.

*From the book *Surgeonfish* by Ingrid Wendt, published in 2005 by WordTech Imprint, Cincinnati, Ohio. Excerpted with permission.

Surgeons often travel in schools, and on a sunlit day it's a cerulean treat for the eyes when hundreds of blue tang sweep across the reef in search of food. The Atlantic Blue Tang is one of the many reef fishes that undergo a brilliant color change as they age. In *A White House Diary* from 1965, the U. S. First Lady at the time, Lady Bird Johnson, recorded her impressions following a snorkeling adventure in the Virgin Islands, among them this one: "I thought the fish were the most wonderful things! There

BORUSIKK / DREAMSTIME

Reaching a meter in length, the Whitemargin Unicornfish (Naso annulatus) is one of about 20 species that sport the distinctive protuberance.

were lots of little bright yellow fish, young blue tang, they called them — as they grow up they change color."

Surgeonfishes are herbivores, guilty of conspicuous consumption as they swoop down upon a verdant patch of algae, grazing *en masse* before moving on. Sometimes, amid this blitz of the buffet table, a smaller fish will be seen darting crazily about, using all of its energy to peck at and chase away the marauders. Pity this poor damselfish trying only to protect an algal patch it has gone to great pains to cultivate. Like a farmer's attempt to chase a tornado out of a carefully tended cornfield, the damsel's effort often appears doomed before the onslaught.

HELMUT CORNELI / ALAMY

A school of meter-long Whitemargin Unicornfish like this one near Tofo, Mozambique, is an impressive sight to behold.

& ASSOC. / ALAMY

AMAR AND ISABELLE GUILLEN/ ALAMY

Atlantic Blue Tang (Acanthurus coeruleus) are yellow as juveniles, blue as adults.

At the intersection of indigenous cultures and reef fishes, surgeons are highly valued as a food source. This is where their ecological importance, a subtle yet important one, plays out in dramatic fashion. When fishing pressure intensifies and reduces populations of surgeons and other reef grazers such as parrotfishes, algae may grow out of control, literally smothering corals. In 1972 marine biologist Sylvia Earle noted that the coral reefs of the Caribbean are "almost devoid of conspicuous plants." Today the reverse is more likely and in many Caribbean islands, once thriving colonies of living corals have been reduced to algae-coated cemeteries.

Further proof of the significance of the surgeons came in 1983 and '84 when disease killed off virtually all Long-spined Sea Urchins in the Caribbean. Urchins are also grazers and another key component in regulating algal "cover" on coral reefs. Fewer surgeons and urchins meant that algal growth went into overdrive, and many reefs in the region have never recovered.

Many surgeonfishes have beguiling names; for example, the Convict Surgeonfish, Lipstick Tang, Two-tone Tang, Bulbnose Unicornfish and the ominous-sounding Razor Sawtail. The Achilles Tang, *Acanthurus achilles*, was named for the hero of Homer's *Iliad,* the Greek warrior slain by an arrow to his heel, his only vulnerable spot. The "heel" of this fish, a brilliant blood-orange spot, must have reminded its namer of the mythical bleeding heel that was Achilles's downfall. But rather than a vulnerability, the orange spot is the site of this fish's nasty scalpel.

A close relative to surgeonfishes is the Moorish Idol, a stunning black-and-gold charmer whose name is apparently linked with African Moors who thought the creature

BLICKWINKEL / TEIGLER / ALAMY

An Achilles Tang (Acanthurus achilles) swims over the coral reef.

a harbinger of good luck and happiness. Moorish Idols are known to pair for life, and what a pleasant thought it is to imagine a bit of finny romance to blunt the edginess in this eat-or-be-eaten environment.

JANE GOULD / ALAMY

A pair of Moorish Idols (Zanclus cornuta) swim over table coral in the Andaman Sea off Thailand.

23
Barbed Wonders

The first time you see a lionfish, you may feel as if you've come upon a contestant from Rio de Janeiro's famed Carnival parade. Outlandish in appearance with brilliant, feathery appendages projecting at all angles, it's a fish that begs to be photographed, an image that speaks to bizarre beauty, the hallmark of so much the coral reef has to offer.

Also known as firefishes, turkeyfishes or zebrafishes, the lionfishes belong to the scorpionfish family, a group of nasties that include some of the most venomous fishes in the sea. The order Scorpaeniformes (skor-peen-ih-FOR-meez) was carefully investigated in the early 19th century by the French naturalist Georges Cuvier, who studied a

A Red Lionfish (Pterois volitans) fearlessly patrols a coral reef in Indonesia. This flashy predator hangs in place in the water or swims casually along like an underwater hovercraft with fins.

DAVID FLEETHAM / ALAMY

diversity of animals (see Chapter 18, for instance) and apparently counted the lionfishes among his favorite creatures to work with. Observations of lionfishes make up one of the earliest understood and best documented natural history studies, reflecting the unbridled fascination of humans with creatures that inflict great pain, or that can do us in with a single bite or sting.

A PRE-DARWIN PIONEER

Besides studying lionfish natural history, Georges Cuvier was also very busy in the fields of paleontology and comparative anatomy. His work with fossils, especially large quadrupeds, led him to pose what seems a quaint understatement today but in his day was rather extraordinary:

"Anyone seen a woolly mammoth lumbering around lately?" he might have asked. "Well, then, since there don't seem to be any mammoths about — anywhere — they must have become extinct!"

Prior to his bold statements on the subject, the very idea of extinction was nonexistent, and until Darwin, no other scientist planted so many seeds of evolutionary thought as did Cuvier.

CHRONICLE / ALAMY

Georges Cuvier lecturing on paleontology at the Museum of Natural History in Paris

Despite their relatively small size — 40 cm long counts as a large specimen — they are one of just a few fishes that allow themselves to be scrutinized at close range, an engaging treat for the diver or snorkeler with camera in hand. A lionfish seems to fear nothing. If challenged, it will aggressively expand its fins and face down the threat. The lionfish's dorsal spine, employed as a spear, can kill marine creatures instantly, and humans who come into contact with the fish are likely to experience intense pain. A lionfish's long, flamboyant and frilly fins are an effective marine billboard that advertises its dangerous nature, and also tricks threatening animals into believing the lionfish is larger than it really is.

Lionfishes' venomous spines serve to confirm one of the basic tenets of nature: Only fools mess with colorful or bizarre-looking creatures that appear fearless. However, there are those who find certain venomous species quite the taste treat; in France the Mediterranean Scorpionfish — *rascasse* — is used to make the fine soup *bouillabaisse.*

Now for the bad news. Lionfishes are turning up in some unexpected quarters, thousands of kilometers from their native habitats among the coral reefs of the Indo-Pacific and the Red Sea. With disturbing frequency they are being spotted in the Atlantic and Caribbean. Recent estimates place over 1,000,000 individuals in the Bahamas alone.

An effective eating machine, the lionfishes are finding such newly invaded areas easy pickings, since their prey have no visceral memory that relays the signal to flee the newcomer. Upwards of 50 species of reef fishes have been found in lionfishes' stomachs, including gobies, wrasses, parrotfishes, sea basses and juvenile groupers and snappers. Size seems to be no deterrent — the lionfish is capable of gulping prey as large as three-fourths its own dimensions.

Let's say you're a hungry lionfish and your day begins in the pre-dawn hours. As you scan a reef — in the Virgin Islands, off Cozumel, in the Florida Keys or around Barbados, what you see is a nice variety of juvenile or adult fishes that don't appear to recognize you as the predator you are. Unlike your Pacific relations do, you don't need to wait for the cover of darkness. Your "cafe" is open 24/7! And so you set out to dine to your heart's content. This event, replicated millions of times, is a recipe for disaster.

DAVID FLEETHAM / ALAMY

This could be the last thing a lionfish's prey will ever see.

SCORPAENOIDS

Spiny-finned fishes, or *scorpaenoids* (SKOR-peen-oids), fall into two groups. The first consists of brightly colored, highly venomous species like the Red Lionfish that helicopter about coral reefs displaying their warning coloration.

And then there are the "cryptic" scorpaenoids, largely confined to the sea bed, whose colors blend with the surrounding environment. As if designed by a mad scientist, these stealth predators are camouflaged with a frightening array of strange flaps and appendages, spines, plates and ridges. The total wardrobe makes them extremely difficult to spot and at times, they look more like a rock or a piece of seaweed than a fish.

ERIK SCHLOGL / ALAMY

As "cryptic" as it gets, the Smallscale Scorpionfish (Scorpaenopsis oxycephala) is one with the surrounding coral reef off Port Moresby, Papua New Guinea.

On many Caribbean reefs, where once the vibrant colors assigned to dozens of reef residents were a thrill for neophyte and experienced divers and snorkelers, reef fish populations are now at great risk. Lionfishes are prolific breeders. A lionfish female can produce 30,000 eggs in a single spawning event and can spawn as often as every 4 days. Annual production can reach more than 2,000,000 eggs per female.

It isn't known with any certainty how and when lionfishes were introduced to the Atlantic and the Caribbean. It's virtually impossible for them to have broken across the African or American continents or to have navigated through the cold waters of South Africa's Cape Horn. So where did they come from? For some time, it was thought they were released by aquarists following Hurricane Andrew in 1992, but there is evidence to suggest they were first observed off Dania Beach in South Florida in 1985.

ANTONIO BUSIELLO / ROBERTHARDING / ALAMY

In their non-native Caribbean waters off Honduras, Red Lionfish are a lovely but unwelcome sight.

Lionfishes are among the reef's major predators. They actively hunt, corner and dine on many tropical species; their diet includes a broad spectrum of reef fishes, and once established, they are very difficult to eradicate. Large groupers and sharks tend to avoid them unless conditioned by scuba divers to accept one from the end of a spear.

But in one of those quirky scenarios that nature throws at us now and then, without having to even get wet, you and I could prove to be helpful combatants against lionfish infestations. It turns out they taste pretty good, and chefs on the lookout for the next new thing are adding them to menus. There are even cookbooks (and recipes to be found on the internet, of course) that give all of us the chance not only to offer something remarkable at our next dinner party, but also to help maintain Caribbean coral reefs in the process.

RON SALLEE

Lionfish on the menu in Willemstad, Curacao

24

Hedgehogs of the Sea

"The Fish found on these coasts, and called the Sea-Urchin, well deserves that name: It is round as a ball, and full of sharp prickles, for which it is feared: Some call it the Armed Fish. They who take of them, having dried them, send them as Presents to the Curious, who for rarity hang them up in their Closets.

— Charles de Rochefort, *History of the Caribby-Islands,* 1658

SEAPHOTOART / ALAMY

Echinothrix diadema, commonly called Diadema or Blue-black Urchin, in Moorea Lagoon, French Polynesia

Sea urchins are *echinoderms,* a word that pairs the Greek *echinos* for spiny and *derma* for skin. The urchin is one more example of how difficult it is to compare a bizarre marine animal to a terrestrial counterpart. A submerged pincushion comes close, and an undersea hedgehog gets closer, but neither begins to capture the mesmerizing view of a sea urchin's spines as they oscillate before the current.

We must take a giant millenial step backwards to uncover the origins of the word *urchin.* In 1066, when the Normans invaded England, their baggage included words unfamiliar to the English. One of them was *herichon,* the word for hedgehog. The English seem to have had difficulty wrapping their tongues around that, so it evolved into "urchin." Because they resembled the terrestrial version, our spiky maritime subjects were once called sea hedgehogs before good sense modified that into sea urchins. Superstition held that hedgehogs were elves or goblins in disguise, and besides naming a sea creature, "urchin" attached itself to mischievous, rag-tag children of the streets.

All sea urchins have a five-part body plan. Circling the oral cavity, the arrangement is defined by the joyous phrase — *pentamerous* (pen-TAM-er-us) *radial symmetry.* They are related to sea cucumbers, sea stars, brittle stars and feather stars and are found clinging to rocks, corals, sponges and sand. Many people may know them best for their painted souvenir skeletons or for the availability of their gonads, known as *uni*, at the sushi bar. Far more than this, urchins are complex, stunningly beautiful creatures whose geometrical features have astounded and inspired not only scientists but artists as well.

The urchin constructs its exoskeleton — called a "test" — by extracting calcium carbonate from the sea. The 19th-century English naturalist, J. G. Wood, captivated by the structure, wrote "…if you have the empty box of an urchin…hold it up to the light, and look into the cavity from the under or mouth side…light streams in through a multitude of minute holes, as smooth and regular as if drilled with a fairy's wimble…"

The five-part radial symmetry characteristic of sea urchins can be seen in this fossil. Fossilized urchins like this one found in England are unusual in their regularity in size and mass. Known colloquially as Chedworth Buns, Checkbury Buns or Poundstones, they were used by Oxfordshire milkmaids as weights for butter scales, a practice that continued into the 18th century.

CARL COFFMAN / DREAMSTIME

Close-up of a sea urchin skeleton, or test, backlit

NOURISHMENT & INSPIRATION

Prominent Spanish surrealist Salvador Dalí loved dining on sea urchins, fresh from Mediterranean waters bordering the tiny Catalonian fishing village of Cadaqués.

ROBERT DESCHARNES / ALAMY

When he wasn't eating them, Dalí used them for inspiration. He was fascinated by the shape of the creature's exoskeleton, a rigid structure that protects the urchin's soft interior. The form appears in much of Dalí's work. In *50 Secrets of Magic Craftsmanship* (1948) he weighs in on the urchin's role in the creative process: "Secret number 45 concerns the aesthetic virtues of... the sea urchin, in which all the magic splendors and virtues of pentagonal geometry are found resolved, a creature weighted with royal gravity and which does not even need a crown for, being a drop held in perfect balance by the surface tension of its liquid, it is world, cupola and crown at one and the same time, hence universe!"

Aristotle, who first described the body structure, habits and diversity of urchins, compared the "mouth apparatus" to a "horn lantern with the panes of horn left out." Horn lanterns of that time were five-sided and constructed of five panes of translucent horn, sufficiently thin to permit light to shine out but substantial enough to protect the candle from the wind. The label "Aristotle's lantern" belies the complexity of the mouth, an intricate bit of marine architecture whose function is dictated by a complex set of muscles and plates, which the animal uses to scrape algae from rocks and other surfaces.

The urchin's prodigious appetite helps keep the reef from being smothered by seaweeds. But the situation can get out of hand. Other than man, not many creatures are capable of making a meal of the urchin. Triggerfishes can, but people eat triggerfishes, sometimes to excess.

NEIL SETCHFIELD / ALAMY

The Aristotle's lantern is the feeding apparatus (mouth and teeth), found in the center of the underside of the sea urchin.

If their predators are depleted, sea urchin populations explode and the urchins chew up every bit of algae in sight. A marauding gang of urchins acts like an army of underwater ants, able to strip the seafloor of all edible plants in their path, and may also consume young coral "recruits" and other invertebrates living there. In some studies, urchin populations were a hundred-fold more dense in areas where there was over-fishing of their predators. After urchins strip an area of algae, other reef grazers — parrotfishes for example, may starve and their numbers plummet. In a balanced ecosystem, in which urchin populations are held in check by predators, other grazers can compete and everyone's content.

CAROLINE S. ROGERS

A group of Long-spined Sea Urchins (Diadema antillarum) graze the shallows of Leinster Bay in Virgin Islands National Park, St. John, U. S. Virgin Islands.

GIMME SHELTER

Besides their role as grazers, long-spined sea urchins play another important part in coral reef ecology. Their spines protect not only their own larvae and young, but also larvae and small adults of other species.

CAROLINE S. ROGERS

This Grass Squid (Pickfordiateuthis pulchella) is tiny enough to slip between the spines of Diadema antillarum to find protection from predators.

FENKIE SUMOLANG / DREAMSTIME

On Indo-Pacific coral reefs the Double-striped Clingfish (Discotrema lineatus) finds refuge among the spines of the Blue-black Sea Urchin (Echinothrix diadema). It may also take a little nourishment there, nibbling the tube feet (numerous tiny appendages operated by hydraulic pressure) that its host, like other echinoderms, uses for locomotion, feeding and respiration.

25
Ocean Asteroids

To consider narrowly all the rarities to be seen in the Sea, it might be said, that of whatever is excellent in the Heavens there is a certain resemblance in the Sea, which is as it were the others looking-glass. Hence it comes, that there are Stars to be seen in it, having five points or beams, somewhat of a yellowish colour.

— Charles de Rochefort, *History of the Caribby-Islands,* 1658

The term *sea star* crept into our vocabulary in 1569 as "a kind of fishe called Stella, or Sea starre, bycause it hath the figure of a painted starre."* Not long after, the word *sterrefyshe* landed in dictionaries, later evolving to starfish. Scientists refer to them as *asteroids,* from the scientific name of this class of animals, Asteroidea, which joins the Greek *aster* for star with *eidos* for form. The terms sea star and starfish are commonly used, but what fun it would be to alert other snorkelers with "Look here, I've found an asteroid!"

*From "Certaine secrete wonders of nature containing a descriptio[n] of sundry strange things, seeming monstrous in our eyes and judgement, bicause we are not privie to the reasons of them" by Edward Fenton, 1569

Of asteroids, there are some 2,000 species. The "basic model" is five-armed, but some may have up to 50. They are found in each of the world's oceans, from tropic shallows to deep, chilly canyons over 6,000 meters down. Sea stars range from fingernail size to waist-encircling giants such as the Sunflower Sea Star at a meter or more from arm-tip to opposite arm-tip. Their colors cover the spectrum — red, purple, orange, blue and yellow are not uncommon. They may be striped,

ETHAN DANIELS / STOCKTREK IMAGES / ALAMY

A Blue Sea Star (Linckia laevigata) on a coral reef in Komodo National Park, Indonesia

polka-dotted, spiny or smooth, svelte or robust, with names like Goosefoot Sea Star, Bloody Henry Sea Star and Chocolate Chip Sea Star.

The combination of form, color and texture has made the starfish an enduring symbol of the sea. We wear them as tattoos and they adorn cocktail napkins, find their way onto key chains and are cast in precious metals as jewelry. The witless starfish Patrick Star is well known as SpongeBob SquarePants's best friend. An asteroid is even tucked into the lunch box of little Cynthia Rose in the musician Prince's catchy tune, "Starfish and Coffee."

In the coral reef fashion show, sea stars would be the dazzling, leggy supermodels. But looks can be deceiving. Despite their beauty and innocent bearing, a sea star is a mobile — albeit slow-paced — rapacious killer that employs a crafty technological advantage to stalk and consume its prey.

"SKIN" IN THE GAME

Cuddly, sea stars are not; their "skin" is a calcified armor against the rare predator (sometimes another starfish) that might wish to eat them. Some are covered with skin-piercing thorns, and still others are toxic. Nor do they particularly enjoy being handled. A physiological quirk known as *mutable connective tissue* allows the sea star to harden its entire body within seconds of being grabbed. When the threat passes, the creature can immediately soften and be on its merry way. Mutable connective tissue, found in all classes of echinoderms, is of interest for its potential to help us understand and possibly treat arthritis.

SUE DALY / NATURE PICTURE LIBRARY / ALAMY

Not a resident of coral reefs, the largest starfish, Sunflower Sea Star (Pycnopodia helianthoides), resides in cooler waters. Since this photo was taken on the Pacific coast of Canada in 2009, there has been a die-off of this species associated with warmer than usual water.

ETHAN DANIELS / STOCKTREK IMAGES / ALAMY

The Chocolate Chip Sea Star (Protoreaster nodosus) lives in warm shallow waters in the Indo-Pacific region.

The sea star uses a sophisticated hydraulic system to expand and contract thousands of tube feet on its underside. The tube feet allow it to clutch and grab its way across the sea bed or to cling to rocky shorelines pounded by surf. At mealtime, "dining out" is taken to the extreme. Instead of transporting food to its digestive system, the sea star does the exact opposite. Coming upon a tasty worm, sponge or bit of algae, our creature everts its stomach from its mouth, enveloping the main course. Digestive enzymes do their thing, the stomach is pulled back in and our starfish goes on its sated way. One asteroid, the Crown-of-Thorns, has such a prodigious appetite for coral polyps that infestations of these sea stars have periodically laid waste vast swaths of Pacific coral reefs.

CAROLINE S. ROGERS

A red Cushion Sea Star (Oreaster reticulatus) consumes a sponge in Hurricane Hole, St. John, U. S. Virgin Islands.

Clams and other bivalves are more challenging prey, and this is when the sea star's tube feet come into play. Enveloping the shell in a deadly embrace and exerting a steady hydraulic pressure through its tube feet, the sea star pulls outward. The clam "clams up" for as long as it can, but eventually it weakens and is pried open. The sea star inserts its stomach, dinner is served and the rest is history.

In some cases asteroids become the prey. Seagulls eat them, and humans consume them on a limited scale. More often, though, the human demand for starfishes is not for food but as dried specimens. Because harvesting them is so easy, they are often overcollected, then sold as cheap souvenirs. Shape intact, their splendid colors long faded, they hang from our Christmas trees, we lean them against the wall on shelves, or we plop them into aquaria. The lucky ones we leave to the sea's embrace.

GARY E. DAVIS

A Crown-of-Thorns starfish (Acanthaster planci) eating coral polyps on the Great Barrier Reef, Queensland, Australia

In more than one version of a story adapted from Loren Eiseley's 1969 essay "The Star Thrower," a man comes across a child throwing marooned sea stars back into the sea. Strong waves have pitched them onto the beach, high above the line of drift seaweed, where they will surely die. "Child," he says, "do you not realize that there are miles and miles of beach, and there are starfish all along every mile. What you're doing can't possibly make a difference!" The child launches yet another sea star beyond the surf, and replies as it splashes into the water, "It made a difference to that one."

LIZA FLORES

A determined child rescues stranded sea stars in this detail from "Star Thrower," an illustration from the book Earth Tales: 3 Eco Fables for Children, *published by CANVAS (Center for Art, New Ventures & Sustainable Development).*

BRAINLESS

Given their success at predation, it's hard to believe that sea stars (starfishes) are brainless creatures. But a starfish navigates life's daily challenges not by a centralized brain but via a nerve network that branches throughout the body. Though incapable of solving quadratic equations, the asteroid is no slouch when it comes to dealing with traumatic events. Most starfishes can regenerate severed limbs, and in a science fiction fantasy come true, certain species can replicate themselves from a single separated limb alone, as long as it has enough of the animal's central disc still attached. Fishermen used to cut sea stars in half, believing they were rescuing a molluscan fishery from the predatory echinoderm. Instead they were adding to the population, as the sliced-up sea stars multiplied.

Looking Forward

PHOTOS: CAROLINE S. ROGERS

When I lived on St. John, in the U. S. Virgin Islands, I liked to snorkel an area at the edge of Francis Bay. Submerged boulders that had long ago tumbled from the steep hillsides created ideal grottoes and caves for sheltering groupers, eels and octopuses. The water was shallow and often calm, and so a place where the residents spooked easily. You needed to be a patient and calm snorkeler if you hoped to see anything. Coral reefs in general most reward those who come in peace.

I enjoyed looking for the juvenile fishes and invertebrates such as sea stars and sea urchins that sheltered on this small reef; tough little residents, but also vulnerable. There were golf-ball-size Boulder and Brain corals as well as tiny Elkhorns and Staghorns. As corals grow, countless reef residents shelter amongst them, some with a high degree of fidelity to place. If you can remember where you saw that Green Moray eel one time, chances are you'll find it there the next.

It was thrilling to see a juvenile Yellowtail Damselfish no bigger than a half-dollar, speckled with iridescent blue spots, a fish that looks like it should be on display in a jewelry shop.

And there were young Queen Angelfish, outfitted in electric blue hues, shouting "Look at me!" Angelfish are irrepressibly curious when fish watchers visit. They play hide and seek, darting out of sight, then suddenly reappearing as if to say "Are you still here?"

Sea fans adorned with Flamingo-Tongue Snails looked to me like flattened Christmas trees decorated with living ornaments.

Here too were Banded Coral Shrimp along with Christmas tree and feather duster worms, invertebrates common to Caribbean reefs, and whose names, like the shrimp and feather duster shown here, leave no doubt as to their appearance.

There were uncommon sights as well. One day I saw a baby boxfish, no larger than the tip of my thumb, furiously beating its fins to stay in one place. I also had the rare privilege to witness a Manta Ray with a wingspan twice that of my own as it "barrel-rolled" several times right underneath me. Once I found myself surrounded by baby octopuses, adrift in the water column, still encased within their eggs, somehow prematurely detached from their protective grotto. Of this dreamlike experience, I still ask myself "Did that really happen?" Seeing such new life is unusual on the coral reef. But something always happens; no two visits are ever the same.

PHOTOS: CAROLINE S. ROGERS

* * *

OK, time for us to take a deep breath and dive on in. It might be a tad unpleasant at first, but stay with me.

As seas warm, corals are stressed and become pale or totally white as they lose the microscopic algae in their tissues. Such bleaching "events" are becoming more frequent, extended and damaging. Also associated with warming seas, coral diseases have increased as stress makes it harder for corals to resist infection.

Warmer seas are also the high-octane fuel that hurricanes feast on. Higher water temperatures turn low-pressure systems into storms with intense and powerful circulation that leads to categorized hurricanes. And like an unwanted guest reluctant to leave, when the water is unusually warm, hurricanes tend to move slowly, gaining in strength as they go. If the intensity and frequency of such storms climbs in years ahead, it will be more difficult for coral reefs and their inhabitants to fully recover.

Today coral reefs are still a repository of amazing biodiversity, still dazzle us when we visit them and are still inspirational. At aquariums and nature parks, creative programs are connecting people to coral reefs, building support for conservation where it means the most — at the level of the curious individual, perhaps especially children. Numerous coral reef restoration projects are underway, and the network of marine protected areas (MPAs) around the world's coral reefs is growing every year. These MPAs are designed to safeguard the reefs from such things as overfishing, pollution and anchor damage. The hope is that such protection will, among other things, result in healthier reefs where corals flourish and fish populations replenish themselves.

At the research level, scientists have observed that when reefs are subjected to hurricanes and other stresses, the effects seem to be "patchy," some areas having avoided the damage and leading the way to recovery. I learned that sometimes hurricanes that come close, but not too close, to coral reefs can even be beneficial because they lower the local water temperature, temporarily at least, reducing thermal stress. Also, in some areas where thermal stress seems likely to be the new normal, certain species of corals appear to be getting better at tolerating high temperature, holding out the tantalizing possibility that their algal symbionts are becoming adapted to warmer seas. In an ambitious program at the Great Barrier Reef, fragments of shallow-water, heat-resistant corals are being transplanted to severely bleached reefs in an effort to aid in the recovery of degraded areas.

All these strategies for coping with rising temperatures are well worth the considerable effort required. But a way to address the underlying problem is right in front of us. From now on we have to minimize the amounts of heat-trapping gases, such as carbon dioxide, that we put into the atmosphere. To date, one of the most effective ways we know of to reduce the rate of increase of atmospheric carbon dioxide is to reduce our use of fossil fuels. With luck, we stand a chance of slowing or reversing the trend toward ever warming seas.

I wrote *Coral Reef Curiosities* because I was smitten with the idea that humans have an attachment to coral reef animals that extends well beyond their watery bounds, surprising connections that I hope have left an impression as you've read about them. One thing I didn't expect to find in the process of writing the book was this: There are heroes among us that you rarely hear about but to whom we owe a great deal. They follow in the footsteps of a long line of dedicated individuals, some whose stories are woven into these pages, many of whom lived in far reaches of the planet under less than ideal conditions. There they collected, named and archived in pictures and words the unusual and the curious, bolstering the library of knowledge we have today. Our modern-day heroes are of a slightly different breed, their cause less tilted toward inventorying new forms of life. Rather than cataloging species, some study how coral reef systems work, sharing knowledge so that others can carry out what is a rescue mission of escalating importance.

Those who live on islands like St. John have what would seem an idyllic life in some of the most beautiful locations the world has to offer — until disaster strikes. With disturbing frequency, they shelter from storms that shred their homes, offices and lives, ripping the heart out of the places they love. You have to live through the shrieking violence of a high-energy storm to understand the terror of that experience. It would be easy for them to abandon their assignments for safer postings or to retire. Who could blame them?

Yet time after time, they draw upon an inner resolve whose bonds to the marine world are unbreakable. After the storm passes, long before the restoration of what most of us would call a normal life, they don mask and fins, strap into scuba gear and return to the coral reefs. They study what's left, document the damage and the recovery, and try to figure out how to move forward so people like me can visit a little pocket of wonder at the edge of an island, there to ponder our collective attachment to the curiosities of a coral reef world.

Del Mar, California
February, 2020

reefcuriosity@gmail.com

Selected References

These are the books, articles and websites — organized here by book chapter — that I found most helpful while doing research for *Coral Reef Curiosities*. Some of the books listed are modern printings of the original references. Discovering the wealth of background material that has been preserved, scanned and posted online was a sheer joy and a revelation to me. Much of this work is rare and virtually inaccessible outside of the "gatekeeper" url's that so often tugged me into the rabbit hole of "research rapture" to which I so willingly subscribed.*

The Blade That Became a Book

Horwitz, Tony. *Midnight Rising: John Brown And The Raid That Sparked The Civil War.* Henry Holt (2011)

1 Sixty Degrees of Separation

Coniff, Richard. "Useless Creatures." *New York Times* (September 13, 2014)

*The url's cited as references are valid at the time of publication. While two things in life are guaranteed, the longevity of a url most likely is not. Please advise the publisher if you come across any discrepancies that prove this point.

Darwin, Charles. *Voyage of the Beagle.* Penguin Books (1989)

Jones, Steve. *Coral: A Pessimist in Paradise.* Little, Brown (2007)

Mitchell, Andrew. *The Fragile South Pacific: An Ecological Odyssey.* University of Texas Press (1990)

Pauly, Daniel. *Darwin's Fishes: An Encyclopedia of Ichthyology, Ecology and Evolution.* Cambridge University Press (2004)

Rogers, Caroline S. *Coral Reef Stars: A Galaxy of Undersea Images.* Caroline S. Rogers (2009)

2 Animal Blossoms

Fautin, Daphne G. and Gerald R. Allen. *Anemone Fishes and Their Host Sea Anemones.* Sea Challengers (1997)

Mariscal, Richard N. *A Field and Laboratory Study of the Symbiotic Behavior of Fishes and Sea Anemones from the Tropical Indo-Pacific.* University of California Press (1970)

Shick, J. Malcolm. *A Functional Biology of Sea Anemones.* Chapman & Hall (1991)

3 Aristotle's Sponges

Aristotle (author); Thompson, D'Arcy Wentworth (translator). *The History of Animals.* D'Arcy Wentworth Thompson (1910)

Bergquist, Patricia R. *Sponges.* University of California Press (1979)

Rützler, Klaus. "Sponge diving — professional but not for profit." In *Methods and Techniques of Underwater Research, Proceedings of the American Academy of Underwater Sciences 16th Annual Diving Symposium.* American Academy of Underwater Sciences (1996)

4 Tale of the Tunicate's Tail

Goodbody, Ivan. "Drugs From the Sea — Harvesting Sea Squirts." In *University of the West Indies. Faculty of Natural Sciences Newsletter Vol. 7. No.1* (November, 1993)

Humann, Paul, Ned DeLoach and Les Wilk. *Reef Creature Identification, Florida Caribbean Bahamas. 3rd Edition (Reef Set)* New World Publications (2013)

Ponsonby, David and Georges Dussart. *The Anatomy of the Sea: Over 600 Creatures of the Deep.* Chronicle Books (2005)

Sulloway, Frank J. *Freud, Biologist of the Mind: Beyond the Psychoanalytic Legend.* Harvard University Press; reprint edition (1992)

5 Sea Slug Fest

Behrens, David W. *Nudibranch Behavior.* New World Publications (2005)

Cobb, Gary and Richard C. Willan. *Undersea Jewels, a Colour Guide to Nudibranchs.* CSIRO Publishing (2006)

Coleman, Neville, Gary Cobb and David Mullins. *Nudibranchs Encyclopedia: Catalogue of Asia/Indo Pacific Sea Slugs.* Masalai Press/Underwater Australia (2015)

Debelius, Helmut and Rudie H. Kuiter. *Nudibranchs of the World.* Hollywood Import & Export (2007)

6 Beguiling Blennies

Carpenter, Kent E. "A short biography of Pieter Bleeker." *The Raffles Bulletin of Zoology 2007, Supplement No. 14* (January 31, 2007)

Marchant, Sylvia. "Denizens of the Deep: Pieter Bleeker's 'Fish Atlas.'" *The National Library Magazine,* National Library of Australia (March, 2011) https://webarchive.nla.gov.au/awa/20120417162935/http://www.nla.gov.au/pub/nlanews/2011/mar11/denizens-of-the-deep.pdf

Merrillees, Scott. *Batavia in Nineteenth Century Photographs.* Routledge (2000)

Patzer, Robert A. (editor) *The Biology of Blennies.* CRC Press (2009)

7 Winged Flyers

Lorenzini, Stefano. *The Curious and Accurate Observations of Mr. Stephen Lorenzini of Florence on the Dissections of the Cramp-Fish.* Gale Ecco, Print Editions (2018)

McDavitt, Matthew T. "The Cultural Significance of Sharks and Rays in Aboriginal Societies Across Australia's Top End" (2005) **www.mesa.edu.au/seaweek2005/pdf_senior/is08.pdf**

Thompson, John M. (editor). *The Journals of Captain John Smith: A Jamestown Biography.* National Geographic Society (2007)

8 Hina's Eel

Andersen, Johannes C. *Myths and Legends of the Polynesians.* Dover Publications (2011)

Bshary, Redouan et al. "Interspecific communicative and coordinated hunting between groupers and giant moray eels in the Red Sea." *PLOS Biology 4(12).* (2006) **https://journals.plos.org/plosbiology/article?id=10.1371/journal.pbio.0040431**

Knowlton, Nancy. *Citizens of the Sea: Wondrous Creatures from the Census of Marine Life.* National Geographic Society (2010)

Mehta, Rita S. and Peter C. Wainright. "Raptorial pharyngeal jaws help moray eels swallow large prey." *Nature 449(7158)* (2007) **https://www.researchgate.net/publications/6035780**

9 Your Place or Mine?

Beebe, William. *Zaca Venture.* Harcourt, Brace (1938)

Gould, Carol Grant. *The Remarkable Life of William Beebe: Explorer and Naturalist.* Island Press (2004)

10 Breathtaking!

Crowley, Terry. *Beach-la-Mar to Bislama: The Emergence of a National Language in Vanuatu.* Clarendon Press (1990)

London, Jack. The *Cruise of the Snark.* MacMillan (1911)

Toral-Granda, Verónica et al. (editors). *Sea Cucumbers: A Global Review of Fisheries and Trade.* Food and Agricultural Organization of the United Nations (2008)

11 The Fish That Fishes

Pietsch, Theodore W. (editor). *Fishes, Crayfishes, and Crabs: Louis Renard's Natural History of the Rarest Curiosities of the Seas of the Indies.* The Johns Hopkins University Press (1995)

Pietsch, Theodore W. and William D. Anderson, Jr. *Collection Building in Ichthyology and Herpetology.* The American Society of Ichthyologists and Herpetologists (1997)

12 Armed and Curious

Anderson, Roland C., Jennifer A. Mather, et al. *Octopus: The Ocean's Intelligent Invertebrate.* Timber Press (2010)

Courage, Katherine Harmon. *Octopus! The Most Mysterious Creature in the Sea.* Current (2013)

Gifford, Edward Winslow (compiler). *Tongan Myths and Tales.* Bernice P. Bishop Museum (1924)

Lane, Frank W. *Kingdom of the Octopus: The Life History of the Cephalopoda.* Sheridan House (1960)

Montgomery, Sy. *The Soul of an Octopus.* Atria Books (2015)

Stubbs, Alexander L. and Christopher W. Stubbs. "Spectral discrimination in color blind animals via chromatic aberration and pupil shape." *PNAS 113, No. 29* (July 19, 2016) **www.pnas.org/content/113/29/8206**

13 Reef Butterflies

Artedi, Peter. *Ichthyologia* (reprint). J. Cramer (1962)

Blunt, Wilfred. *Linnaeus, The Compleat Naturalist.* Frances Lincoln Ltd. (2001)

Burgess, Warren E. *Butterflyfishes of the World.* TFH Publications (1978)

Pietsch, Theodore W. *The Curious Death of Peter Artedi.* Scott & Nix, Inc. (2010)

Seba, Albertus. *Cabinet of Natural Curiosities.* Taschen (2016)

Wheeler, Alwyne. "Peter Artedi, founder of modern ichthyology." *Proceedings V Congress of European ichthyologists, Stockholm, 1985.* Swedish Museum of Natural History (1987)

14 Smile!

Delbourgo, James. *Collecting the World; The Life and Curiosity of Hans Sloane* (reprint). Harvard University Press (2019)

Gudger, Eugene W. "Sphyraena barracuda: its morphology, habits and history." *Tortugas Laboratory Papers 12(4).* Department of Marine Biology of the Carnegie Institution of Washington (1918)

de Rochefort, Charles. *The History of the Caribby-Islands.* (reprint of the original edition of 1666). Hansebooks (2017) **http://archive.org/details/historyofcaribby00roch**

de Sylva, Donald P. *Systematics and Life History of the Great Barracuda.* University of Miami Press (1963, second edition 1970) **http://scholarlyrepository.miami.edu/cgi/viewcontent.cgi?article=1014&context=trop ocean**

Stearns, Raymond Phineas. "James Petiver: Promoter of Natural Science, c.1663-1718," in American Antiquarian Society, *Proceedings 62* (October, 1952) **https://www.americanantiquarian.org/proceedings/44807240pdf**

15 The Perils of Exquisiteness

Carr, Archie. *So Excellent a Fishe: A Natural History of Sea Turtles*. Scribner's, for The American Museum of Natural History (1967)

Luomala, Katharine. *Voices on the Wind: Polynesian Myths and Chants.* Bishop Museum Press (1955, 1986)

Safina, Carl. *Voyage of the Turtle: In Pursuit of the Earth's Last Dinosaur.* Holt Paperbacks (2007)

Spotila, James R. *Sea Turtles: A Complete Guide to Their Biology, Behavior, and Conservation.* The Johns Hopkins University Press (2004)

Witherington, Blair. *Sea Turtles: An Extraordinary Natural History of Some Uncommon Turtles.* Voyageur Press (2006)

16 The Long Arms of the Maw

Ellis, Richard. *The Search for the Giant Squid.* Lyons Press. (1998)

Palumbi, Stephen R. and Anthony R. Palumbi. *Extreme Life of the Sea.* Princeton University Press (2015)

Prager, Ellen. *Sex, Drugs, and Sea Slime: The Oceans' Oddest Creatures and Why They Matter.* University of Chicago Press. (2012)

Williams, Wendy. *Kraken: The Curious, Exciting, and Slightly Disturbing Science of Squid.* Harry N. Abrams (2011)

17 "A Marveilous Straunge Fishe"

Castro, José I. "On the origins of the Spanish word 'tiburón', and the English word 'shark.'" In *Environmental Biology of Fishes 65:3* (2002)

Compagno, Leonard, Marc Dando and Sarah Fowler. *Sharks of the World.* Princeton University Press (2005)

Crawford, Dean. *Shark*. Reaktion Books (2008)

Eilperin, Juliet. *Demon Fish: Travels Through the Hidden World of Sharks.* Pantheon (2011)

18 A Reef Fish That Does Headstands

Bloch, Marcus Elieser. *Ichthyologie ou Histoire Naturelle des Poissons.* Text-only reproduction of the original 1787 work. BiblioLife (2009) Also (with illustrations): **https://archive.org/details/IchtyologieouHi00BlocA**

Humann, Paul and Ned DeLoach. *Reef Fish Identification: Florida Caribbean Bahamas, 4th Edition (Reef Set).* New World Publications (2014)

19 You Laugh, Vex Chief, He Break Your Head

Coulter, John. *Adventures in the Pacific.* (Classic Reprint edition of the original 1845 work) Wentworth Press (2012)

Druett, Joan. *Rough Medicine: Surgeons at Sea in the Age of Sail.* Routledge (2000)

Gudger, E. W. "Puffer fishes: some interesting uses of their skins." *Scientific American* (April 1920) http://books.google.com/books?id=Pfzg2F9L024C&pg=PA321&lpg=PA321&dq

Halstead, Bruce W. *Poisonous and Venomous Marine Animals of the World, Edition 2.* The Darwin Press, Inc. (1988)

20 Queer Fish

Heemstra, Phillip C. and John E. Randall. *FAO Species Catalogue Groupers of the World.* Food and Agriculture Organization of the United Nations (1993) http://www.oads.org.br/livros/36.pdf

Miall, L. C. *The Early Naturalists: Their Lives and Work (1530–1789)* (facsimile reprint of original 1912 edition). Kessinger Publishing (2010)

Paepke, Hans-Joachim *Bloch's Fish Collection in the Museum Für Naturkunde der Humboldt Universität zu Berlin.* A. R. G. Gantner (1999)

21 Super Males

Hansen, Thorkild. *Arabia Felix: The Danish Expedition of 1761–1767.* Harper and Row (1964)

Maempel, George Zammit. "The Arabian Voyage 1761–67 and Malta: Forsskål and his contribution to the study of local natural history." In *Malta Historical Society, Proceedings of History Week 1992* (1994) http://melitensiawth.com/incoming/Index/Proceedings%20of%20History%20Week/PHW%201992/03s.pdf

Rajan, P. T. *Guide to Chaetodontidae (Butterfly Fishes) and Scaridae (Parrot Fshes) of the Andaman and Nicobar Islands.* Zoological Survey of India (2010) http://archive.org/details/dli.zoological.hpg.048/page/n1

Rautiala, Marjatta "Family Background of Peter Forsskål, Linnaean Disciple born in Finland." In *The Linnean 27(1)* (2011) http://docplayer.net/55952548-Newsletter-and-proceedings-of-the-linnean-society-of-london.html

Streelman, J. T. et al. "Evolutionary History of the Parrotfishes: Biogeography, Ecomorphology, and Comparative Diversity." In *Evolution 56(5)* (2002)

22 Surgeons of the Sea

Carcasson, R. H. *A Field Guide to the Coral Reef Fishes of the Indian and West Pacific Oceans.* Wm. Collins & Sons. (1977)

Randall, John E. *Surgeonfishes of the World* (*Bishop Museum Bulletin in Zoology* series). Mutual Publishing (2002)

23 Barbed Wonders

Coleman, William. *Georges Cuvier, Zoologist: A Study in the History of Evolutionary Theory.* Harvard University Press (1964)

Cuvier, Georges (Theodore W. Pietsch, editor). *Historical Portrait of the Progress of Ichthyology, from its Origins to Our Own Time.* The Johns Hopkins University Press. (1995)

Stott, Rebecca. *Darwin's Ghosts: The Secret History of Evolution.* (Reprint of the original 2013 edition) Spiegel & Grau. (2012)

24 Hedgehogs of the Sea

Allen, Gerald R. and Roger Steene. *Indo-Pacific Coral Reef Field Guide.* Sea Challengers (1998)

Dalí, Salvador. *50 Secrets of Magic Craftsmanship.* The Dial Press (1948)

Voultsiadou, Eleni and Chariton Chintiroglou. "Aristotle's lantern in echinoderms: an ancient riddle," in *Cahiers de Biologie Marine 49(3).* Aristotle University (2008) **http://www.researchgate.net/publication/289275888_Aristotle's_lantern_in_echinoderms_An_ancient_riddle**

25 Ocean Asteroids

Cole, Brandon. *Reef Life: A Guide to Tropical Marine Life.* Firefly Books (2013)

Hendler, Gordon et al. *Sea Stars, Sea Urchins, and Allies.* Smithsonian Institution (1995)

Kaplan, Eugene H. *Sensuous Seas: Tales of a Marine Biologist.* Princeton University Press (2006)

Looking Forward

Berkelmans, Ray, and Madeleine van Oppen. "The role of zooxanthellae in the thermal tolerance of corals: a 'nugget of hope' for coral reefs in an era of climate change." *Proceedings of the Royal Society B: Biologial Sciences 273* (2006) **https://www.researchgate.net/publication/6858946**

"Biggest coral reseeding project launches on Great Barrier Reef." *Phys.org* (2018) **https://phys.org/news/2018-11-biggest-coral-reseeding-great-barrier.html**

"Marine Protected Areas of the U. S. Virgin Islands." *Reef Connect.* **https://www.reefconnect.org/marine-protected-areas/**

Rogers, Caroline S, "Unique Coral Community in the Mangroves of Hurricane Hole." U. S. National Park Service. (2017) **https://www.nps.gov/articles/coral-community-in-mangroves.htm**

Acknowledgments

This book is about magical animals that live in magical places. I wrote it for those who have been to these places — and those who have not, but who nevertheless might be inspired to support living communities of species other than our own, and in doing so, to support ourselves. But I could not have written it on my own. So many people offered their time and thoughts as I wrote this book that it isn't possible to thank them all. A few stand out for their remarkable kindness and for providing invaluable assistance.

I owe special thanks to my friend Caroline Rogers, a renowned coral reef ecologist who is smart, dedicated, generous, an amazing writer and a marvelous underwater photographer. She patiently shared her knowledge of and love for coral reefs during our years together at Virgin Islands National Park. Caroline has helped and encouraged my writing from the beginning and was the first to endorse the idea of this book, offering editorial advice throughout and kindly allowing me to use her beautiful photographs as illustrations.

My editor and publisher, Linnea Dayton, stepped in at the exact moment when I was ready to give up on the idea of this book ever coming to fruition. *Coral Reef Curiosities* wouldn't be what it is if not for her perceptive edits, advice, knowledge and ability to craft an eye-appealing product.

The writer Lisa Fugard was unfailingly supportive with her edits and with keeping the sometimes flickering idea of this book alive. Her experienced eye was crucial to the success of these pages. At last, Lisa, the book you imagined reading in your tropical island hammock after a snorkel on the coral reef is here. Now get thee to that hammock!

Lisa's talented writers' group, my first draft readers, have been unstinting with their time and ever helpful with comments and support. Thanks go to Meenakshi Chakraverti, Linnea Dayton, Mary Frumkin, Mark Radoff, Sheila Sharpe and Nancy Tomich.

For his friendship and early support of my work, I thank the poet and conservationist Ed Zahniser, retired Senior Writer and Editor with the National Park Service.

Ned DeLoach was generous in allowing use of his beautiful photography in this book. Over the years Ned and Anna De Loach, with Paul Humann, have been a fountain of knowledge and inspiration through their beautiful publications in magazines, guide books and documentaries.

I wish to thank the many photographers and other artists whose work is included in this book. Their names appear beside their works and in the "Artists and Photographers" section that begins on page 121.

Ben Frable, the Collection Manager of Marine Vertebrates at Scripps Institution of Oceanography, provided editorial advice and valuable help with species names and characteristics as well as sharing his knowledge of elusive historical references.

I am grateful to Gary Davis, Alan Friedlander, Lisa Fugard, Elizabeth Gladfelter, Nancy Knowlton, Rita Mehta, Sy Montgomery, Candace Oviatt, Caroline Rogers, Helene Rogers Smart, and Ed Zahniser for reviewing and commenting on the manuscript.

My late friend and fellow Tongan Peace Corps Volunteer, David Greenman, was supportive of this book from the beginning and put up with my peregrinations on many a far-flung coral reef. 'Ofa atu, Tevita.

Most of all, I thank my wife, former mermaid and first reader, Rosemary Love, to whom this book is dedicated. She has helped from first to final chapter with editorial comments and unwavering encouragement and has been resolute in her belief that I had something unique to say about coral reef animals that was entertaining and of value. Her partnership with me in life's grand adventure has enabled countless visits to some of this world's most spectacular coral reefs.

Artists and Photographers

Listed here are the photographers and other artists whose work appears in this book, along with information you can use to look for more work by or information about the same artist.*

Alex Mustard alamy.com/EG49DT • naturepl.com

Amar and Isabelle Guillen guillenphoto.com • alamy.com

André Duterte (1753–1842) France

Andy Chia dreamstime.com/andidream_info

Antonio Busiello antoniobusiello.com,
• antonio@antoniobusiello.com • alamy.com

Antonio Tempesta (1555–1630) Italy

Barry B. Brown wildhorizons.com/wp/stock-photography

BC Patch LLC bcpatch.com

Blickwinkel / Teigler fishbase.se/photos/PhotosList.php?id=1102&vCollaborator=Frank+Teigler
• alamy.com/DAGNE7

Borusikk dreamstime.com/borusikk_info

Brian Lasenby brianlasenby.com
• shutterstock.com/g/brianlasenby

*For alamy.com listings, once you arrive at the web page, click on the Photographer's or Contributor's name that appears beneath the photo.

Carl Coffman dreamstime.com/calart_info

Carol Buchanan dreamstime.com/cbpix_info

Caroline S. Rogers coralreefstars.com
• coralreefstars@gmail.com

Cayman Islands Postal Service caymanpost.gov.ky/

CBImages alamy.com/E7M4XX

Chris Heller alamy.com/BM3D86

Chronicle alamy.com/G38DYE

Cigdem Sean Cooper dreamstime.com/lilithlita_info

Claude Lorrain (1600–1682) France

Damsea shutterstock.com/g/damsea

David Fleetham davidfleetham.com
• alamy.com/A0N8J4

Divedog shutterstock.com/g/divedog
• stock.adobe.com/contributor/205020093/divedog

Durden Images shutterstock.com/g/durdenimages

Erik Schlogl redbubble.com/people/eschlogl
• alamy.com

Ethan Daniels / Stocktrek Images oceanstockimages.com
• alamy.com

Fenkie Sumolang dreamstime.com/fenkieandreas_info
• 123rf.com/profile_fenkieandreas

Frantisekhojdysz shutterstock.com/g/frantisekhojdysz

F. Schneider / Arco Images GmbH alamy.com/G9RHP4
- facebook.com/pg/arcoimages/about/

Gary Bell oceanwideimages.com
- facebook.com/GaryBellPhotography

Gary E. Davis gedavis204@gmail.com
- gedapix.com

George Grall nationalgeographic.com/contributors/g/photographer-george-grall
- alamy.com/CFG360

Hans Gert Broeder BunakenHans.com
- dreamstime.com/hgb700_info

Helmut Corneli alamy.com/PD9DFW

Henner Damke dreamstime.com/hdamke_info

Howard Chew alamy.com/H61TKB

Hubert Yann alamy.com/F0BMC0

Hudson Fleece alamy.com/JJH10B

Images & Stories alamy.com/A83W9X

Jane Gould alamy.com/CC5JDW

Jeff Miller jephmiller@mac.com

Jeremy Brown dreamstime.com/jeremykeithbrown_info

J'nel shutterstock.com/g/j'nel

John Anderson johnandersonphoto.com
- dreamstime.com/johnandersonphoto_info
- depositphotos.com/portfolio-1729435.html

Jonathan Churchill shutterstock.com/g/Jonathan+Churchill

Jonathan Lavan underpressurephotog.com

Jurgen Freund naturepl.com/search?s=jurgen+freund
- jurgenfreund.com

Kleberpicui depositphotos.com/stock-photos/turtle-hatchlings-kleberpicu.html

Lano Lan bylanolan.wordpress.com
- shutterstock.com/g/Lano+Lan

Leslie Nawirridj kunwinjku-aboriginal-art.com

Liza Flores lizaflores@gmail.com
- CANVAS.ph

Luca Vaime shutterstock.com/g/lucavaime
- underwatertribe.com

Mabel Dwight (1875–1955) art.famsf.org/mabel-dwight/

Mark Richards markrichards.eu

Matthew Banks alamy.com/DRY1A5

Melvinlee shutterstock.com/g/melvinlee
- dreamstime.com/melvinlee_info

MShieldsPhotos shutterstock/g/mshieldsphoto
- alamy.com/DRH8RD

Ned DeLoach divephotoguide.com/user/neddeloach
- fishid.com/gallery/ned

Neil Setchfield alamy.com/ER55YA

Nicescene shutterstock.com/g/nicescene

Norbert Wu norbertwu.com
- mindenpictures.com/search?s=norbert+wu

Olga Nosova dreamstime.com/olganosova_info

Orlandin dreamstime.com/orlandin_info

Paul Springett A alamy.com/A9RWNA

Peter Leahy dreamstime.com/pipehorse_info

Pics516 dreamstime.com/pics516_info

Pierre Denys de Montfort (1766–1820) France

Pietro Tonelli (early 20th century) Italy

Poelzer Wolfgang alamy.com/BHNXP5
- facebook.com/underwaterphotos.of.the.world/

Reinhard Dirscherl ocean-photo.de
- alamy.com/CEG22R

Rita Mehta and Candi Stafford mehta.eeb.ucsc.edu

Rob Waara rob_waara@nps.gov

Robert Descharnes alamy.com/APFK80

Ron Sallee machias.org/lion-fish--its-on-the-menu.aspx
- ronsallee@frontier.com

Ross / Tom Stack & Assoc. alamy.com/BPYWDB
- tomstackassociates.photoshelter.com

R. T. Pritchett (1828–1907) England

Samuel Fallours (late 17th–early 18th century) Netherlands

Scubazoo scubazooimages.com • alamy.com/ D9YCH0

Seadam dreamstime.com/seadam_info

Seaphotoart alamy.com/JA6GG4

Sergio Llaguno dreamstime.com/sergiollag1_info

S.Rohrlach istockphoto.com /portfolio/
- rohrlach?assettype=image&sort=best

Stephankerkhofs dreamstime.com/stephankerkhofs_info

Stephen Frink stephenfrink.com • alamy.com/A4C390

Steve Jones stocktrekimages.com/results.asp?fotid=SJN
- millionfish.com

Sue Daly/Nature Picture Library alamy.com/E45PBD
- naturepl.com/search?s=sue+daly

Susan Heller heller760@gmail.com

Suwat Sirivutcharungchit dreamstime.com /suwatsir_info

Tara Bonvillain taramariebonvillain@gmail.com
- islandculturearchivalsupport.wordpress.com

The History Collection alamy.com/J7R2YH

Tignogartnahc dreamstime.com/tignogartnahc_info

Tobias Bernhard Raff biosphoto.com/recherche.php?lib=tobias+bernhard+raff

Waterframe alamy.com/B9RWMR

Whitcomberd dreamstime.com/whitcomberd_info

Wim van Egmond sciencephoto.com/contributor/wim/

Index

C

D

E

F

G

P

Q

R

S

U

V

W

X

Y

Z

About the Author

Chuck Weikert was born, raised and educated in northern Delaware, where he lived near a small brook and a wetland known simply as "the swamp." He turned over rocks just to see who lived there. He fished, built dams and tree houses, and captured frogs, toads, snakes, turtles, newts and whatever else showed a willingness to be trotted home to his terrarium. He raised and bred tropical fish, built an ant farm, and one time, managed to release hundreds of baby praying mantises into the house.

At age 10, on a ranger-led tidepool encounter at Acadia National Park, one look at the park ranger was all it took for Chuck to know what he wanted to be when he grew up. Other early influencers were Tom Swift, Superman and fishing magazines, from the realm of reading. Television brought him Curt Gowdy in *The American Sportsman*; Lloyd Bridges as Mike Nelson, freelance SCUBA diver and former U. S. Navy frogman in *Sea Hunt*; and later Jacques Cousteau, marine conservation pioneer.

Not a particularly dedicated student in his youth, Chuck was more interested in learning how to tie flies, read a stream, swim, cycle, play ice hockey, and kick, hit and toss balls. "Somehow I stumbled through college," he says, emerging with SCUBA diver certification and a bachelor's degree in biology from the University of Delaware, ready to see the world.

For two years he was a Peace Corps Volunteer teacher in the Kingdom of Tonga, where he learned not only the local Polynesian dialect but also the "language" of a diverse Pacific coral reef ecosystem. Thus began a lifetime of exploring reefs.

During his career as a U. S. National Park Service ranger naturalist, he spent 13 years at Virgin Islands National Park. As the Chief of Interpretation, he presented programs and wrote extensively about coral reef ecosystems, championing the need to protect and preserve them. His writing has been featured in numerous National Park Service publications and in magazines and newsletters, in print and online.

Now retired, he lives in Del Mar, California with his wife Rosemary and two cats. He volunteers at the Birch Aquarium at Scripps Institution of Oceanography in San Diego, interpreting marine ecosystems and the need to conserve them. When not in front of a spellbound audience at the aquarium's kelp tank, Chuck is as likely to be found haunting the great museums and galleries of the world as he is cruising the waters of a coral reef, always on the hunt for the curious, little known stories, artifacts and human endeavors that unlock the reef's mysteries.

ROSEMARY LOVE

Made in United States
Orlando, FL
12 November 2023